Praise for *The Age-Proof Brain*

"Dr. Marc Milstein's *The Age-Proof Brain* is an entertaining, forward-thinking, and actionable guide to better brain health! By following his practical, easy-to-follow recommendations, you can reduce stress, improve memory, sleep soundly, and set yourself up to live a long, fulfilling life."

—Dr. Edith Eger, *New York Times* bestselling author of *The Choice* and *The Gift*

"I tried many other options, but it wasn't until Dr. Marc Milstein's powerful tips and strategies that I finally learned how to get a great and restful night's sleep. I am grateful that he transformed my habits to protect my brain and help me feel my best. I am so excited that *The Age-Proof Brain* is filled with these same tips and strategies to help others."

—Andra Day, Grammy- and Golden Globe–winning singer, songwriter, and actress

"There's so much you can do today to improve brain health and reduce risk of cognitive decline tomorrow. *The Age-Proof Brain* makes cutting-edge neuroscience accessible to everyone and is a vital guidebook highlighting the critical lifestyle factors that empower you to maintain a sharp, healthy, active brain and protect against dementia."

—Richard Isaacson, MD, neurologist, director of Alzheimer's Prevention Clinic at Florida Atlantic University, and founder of the Alzheimer's Prevention Clinic at New York-Presbyterian/Weill Cornell Medical Center

"Dr. Marc Milstein has the amazing ability to take complicated scientific and medical material and make it easily understandable. He has spoken frequently at our Medical Grand Rounds. Our physicians and staff have learned so much from his cutting-edge updates and applied this to patient care. *The Age-Proof Brain* is a wonderful guide for healthy living."

—Eric Leibovitch, MD, internist, former head of internal medicine at Ventura County Hospital, affiliated with UCLA Medical Center and Providence Hospitals

"*The Age-Proof Brain* is a goldmine of information on brain research with direct application to your lifelong well-being. Dr. Marc Milstein's incredible ability to communicate complex science in simple (and often humorous) examples, case histories, and 'how-to' guidelines is guaranteed to change your life. I have been a sleep researcher/educator for more than 50 years and have never read a better and more accurate chapter on sleep research and practical sleep hygiene than you'll find in this book. As the author states, 'Sleep is perhaps the greatest ally in the fight to preserve your brain.' Great advice in an absolutely fabulous invaluable read!"

—Dr. James B. Maas, Weiss Presidential Fellow, former professor and chair of psychology, Cornell University

"This is a wonderful, life-changing book. You learn how to memorize new information, think with greater clarity, unlock your ability to influence others, and feel wonderful about yourself. This book can change your life in a very positive way."

—Brian Tracy, bestselling author of *Eat That Frog!* and *No Excuses! The Power of Self-Discipline*

"Dr. Marc Milstein has been a very valuable resource for myself and the athletes I represent to maximize performance by optimizing sleep, stress, and overall brain health. *The Age-Proof Brain* is filled with great tips to reach your highest potential each day."

—Pat Brisson, National Hockey League Players' Association agent and cohead of the Hockey Division of Creative Artists Agency

"Memory, blood sugar, mood, energy, immune and gut health—Dr. Marc Milstein illuminates how it's all connected. By taking care of your body, you take care of your brain. *The Age-Proof Brain* is a riveting road map based on cutting-edge research to get the best out of your brain and body today and decades from now."

—Max Lugavere, *New York Times* bestselling author of *Genius Foods*

"*The Age-Proof Brain* is for you if you are looking for practical ways to keep your brain sharp and healthy for as long as possible. Dr. Marc Milstein presents the latest research proving how important the brain–body connection is for preserving and enhancing brain health. No magic pill here—just a hopeful handbook full of practical advice, surprising facts, and real-life stories meant to help you protect your most valuable asset—your mind!"

—Ken Blanchard, coauthor of *The New One Minute Manager®* and *Simple Truths of Leadership*

"An in-depth guide to the latest techniques for boosting late-life brain health. With forward-thinking research on everything from Alzheimer's to the immune system, Dr. Marc Milstein covers a panoply of brain-related issues with humor and panache."

—Nir Eyal, national bestselling author of *Indistractable*

"*The Age-Proof Brain* distills the most current neuroscience research into accessible, fun, and compulsively readable chapters and debunks many of the alarmingly widespread misconceptions about our brains."

—Michelle Gielan, national bestselling author of *Broadcasting Happiness*

THE AGE-PROOF BRAIN

THE AGE-PROOF BRAIN

New Strategies to Improve Memory, Protect Immunity & Fight Off Dementia

MARC MILSTEIN, PhD

BenBella Books, Inc.
Dallas, TX

Illustrations by L.L.M.M.N.Alwis

BenBella Books, Inc.
10440 N. Central Expressway
Suite 800
Dallas, TX 75231
benbellabooks.com
Send feedback to feedback@benbellabooks.com

BenBella is a federally registered trademark.

Printed in the United States of America
10 9 8 7 6 5 4 3 2 1

Library of Congress Control Number: 2022014677
ISBN 9781637741429 (hardcover)
eISBN 9781637741436

Editing by Claire Schulz and Alyn Wallace
Copyediting by Judy Myers
Proofreading by Jenny Bridges and Isabelle Rubio
Indexing by Elise Hess
Text design and composition by PerfecType
Cover design by Derek Thornton / Notch Design
Cover image © Shutterstock / David M. Schrader (sheet music) / Nicku (body anatomy and foot anatomy) / Miguel Zagran (eyes) / Lucky Business (teeth) / D. Kucharski K. Kucharska (cell) / Mrspopman1985 (geometric brain) / pashabo (pattern) / ALEXSTAND (hair) / Ye.Maltsev (algae) / first vector trend (anatomy) / jaojormami (pores) / Christoph Burgstedt (neurons) / Master1305 (skin) / r.classen (brain) / create jobs 51 (xray) / eranicle (bone) / BLACKDAY (face) / komkrit Preechachanwate (leaf) and © 123RF.com / someoneice (red blood cells)
Printed by Lake Book Manufacturing

This book is dedicated to my wife, Lauren, for endless reasons.

CONTENTS

PART III: AGE-PROOF YOUR BRAIN

Source citations for this book are available online at https://drmarcmilstein.com/APBnotes.

INTRODUCTION

AFTER I GAVE A TALK ABOUT THE LATEST SCIENCE ON HOW TO OPTI-mize brain function, a man in the audience approached me. He told me he was taking a pill he'd seen in a TV ad, which claimed it would improve his memory.

"It's absolutely amazing," he said.

"What's the name of the pill?" I asked.

"I can't remember," he replied.

That just about sums up how effective those "magic" pills are. Yes, it's understandable that we're desperate for a quick fix to one of the greatest collective health fears we have: Losing your memory and having your mind go from a high-functioning piece of biological wizardry to a lump of gray matter that struggles to remember what day it is.

When I say "wizardry," I'm not exaggerating. A healthy brain is an amazing thing. It can effectively learn, remember, manage emotions, analyze, make good decisions, create, and innovate. A healthy brain can also let you do the electric slide, laugh (at people doing the electric slide), navigate, and hit a perfect golf shot (well, sometimes). A functioning brain is essential for being the best version of you.

Our brains do change as we age. And our mental functioning changes, too. A youthful brain is more focused, can learn and retain more information, and is more resilient to many conditions. The average human brain shrinks by approximately 5 percent per decade after the age of forty. As you might imagine, a shrinking and shriveling brain can have a devastating impact on memory, focus, and productivity. What's more, brain disorders like Alzheimer's disease and mental health issues are on the rise. Many of us have witnessed the devastation of dementia and depression and have seen how they can rob our loved ones of being *themselves*.

Worldwide, we're at a tipping point in terms of brain health. Consider these alarming stats about our day-to-day brain and mental health:

- In 2020, fifty-four million people worldwide had Alzheimer's disease or other dementias—a 144-percent increase in the last thirty years.[1]
- In the US, 12–18 percent of adults over the age of sixty suffer from memory loss or cognitive problems.[2]
- Among individuals with dementia, those who had a mental disorder such as anxiety or depression developed dementia around five and half years earlier than those without a mental disorder.[3]
- Anxiety disorders—the most common mental illness in the United States—affect 18.1 percent of the US population. That's forty million adults.[4]
- Depression is the number one cause of disability in the world. In the United States alone, sixteen million people (6.7 percent of the US population) suffer from depression.[5]
- Approximately one-third of people in the US who pass away after age sixty-five will die with some form of dementia.[6]
- Fifty-two percent of all workers are experiencing high levels of stress, mental exhaustion, and burnout.[7]

How can we stem the tide of these startling and terrifying trends in brain health? The pills, games and brainteasers, and "secret solutions" that people spend millions of dollars on—none of them will turn an aging brain into a healthy, youthful one. And these trends have profound implications. In our lifetimes, virtually every one of us will witness the effects of an aging brain, whether in ourselves or in someone we love. For me, I first saw it in my grandmother, Bubbi Lorraine.

How I Came to Care About Brain Health

Bubbi Lorraine was laugh-out-loud funny, a professional artist, and a natural athlete. She could make an Eggo waffle taste better than any waffle I'd ever had. As a teenager, I watched that charismatic person slowly disappear as she began to suffer from dementia. In those days, doctors couldn't even tell us what type of dementia she had. Was her dementia due to Alzheimer's? Was it vascular? The doctors told us there was no point in looking further because there was nothing they could do.

Bubbi Lorraine's decline wasn't the only experience that got me interested in finding out more about the body. As someone who struggled with Crohn's disease since I was nine, I witnessed the difficulty practitioners had diagnosing my condition, and I became curious about how the body works. I started doing research in a genetics lab at the Salk Institute every day after school beginning at age fifteen. At UCLA, I performed over a decade of lab research on topics from infectious disease to breast cancer to brain science. And I started to notice a pattern.

In the last few decades, science and medicine have become very specialized, which can be wonderful in many ways. For instance, incredible discoveries from penicillin to gene editing to futuristic advances in surgery are scientific marvels born out of that intense focus. But that specialization has also caused a lack of perspective about the big picture; people tend to think of the body as a series of silos, isolating one system, organ, or process from all the others and making an absolute statement about what may be causing one specific symptom. But as we learn more about each body system, we understand they are actually very much connected. *What happens in one part of the body strongly impacts another.* The more I researched the latest studies, I started to have a clearer and more nuanced picture of how interconnected multiple aspects of our health truly are.

After completing my doctorate, I had an opportunity at UCLA to give a talk on science to "non-scientists." I hated the way science was typically taught—boring, clinical, hard to understand. I wanted to share breakthrough research in a fun, accurate, and compelling way. That one talk led to another, and then another. I saw the powerful impact of delivering the latest research to people in an understandable way and how the talks positively and dramatically changed lives.

As I traveled the world giving talks on brain health, I was surprised to see that the questions people asked and the topics they wanted to hear about were the same in every country I visited. Across all cultures, generations, and economic levels, the same question was asked in a variety of languages: How do I live an optimal life? *The Age-Proof Brain* answers this question.

In This Book

The good news is, serious mental decline is not an inevitable part of aging. For example, it is estimated that a third of all the current cases of dementia could have been prevented by changes in lifestyle.[8] Day-to-day functioning, productivity, memory, and focus can all be maintained and strengthened throughout life. More and more research points us to a compelling conclusion and powerful message: that we—not our genes—can have control over our cognitive health today and the destiny of our brains. In fact, research presented at the 2019 Alzheimer's Association International Conference found that certain lifestyle factors have a greater impact than your genes do on whether you'll develop memory-related diseases. In 2020, researchers at Rush Medical Center in Chicago found that those who followed recommendations for simple lifestyle modifications *reduced their risk of developing Alzheimer's by nearly 60 percent.*[9]

The keys to a healthy, youthful brain lie in the brain's connections with the rest of the body. There are eleven key factors to preserving and enhancing brain health: a robust immune system; a healthy heart; preventing or treating diabetes; treating mental health conditions; quality sleep; a nutritious diet; regular exercise; stress management; avoiding environmental toxins; social connection; and continually learning new skills.

It is the nuances in these factors that may surprise you! For instance, sleep quality is more about getting effective sleep than just getting a certain number of hours a night. When it comes to exercise, we need to rethink the old saying "No pain, no gain." And enhancing your immune system isn't about *boosting* your immune response—in fact, that could do you more harm than good.

And it's not about one of these factors in isolation but the synergistic aspect of how they work together. The combination of these factors is critical for slowing down the negative effects of brain aging, realizing your brain's true potential each day, and protecting your brain from conditions such as dementia, depression, and anxiety. (You may wonder what mental health has to do with brain aging; as we'll explore later, if they're left untreated, depression and anxiety can speed up brain aging.) This is why I am dedicating a chapter to each of these key factors, filled with take-home, actionable tips.

You don't need to worry about the dizzying fads, sound bites, blurbs, and tweets that are often filled with misinformation. In this book, I carefully examine decades of research to distill the key take-home messages into accurate, actionable, and transformative habits you can keep.

This book is broken into three parts.

Part one offers a crash course on your amazing brain, how it develops and functions, and its powerful connections to your immune system, heart, and gut.

Part two examines what surprising and often overlooked factors age the brain.

Part three provides actionable solutions to keep your brain young and age proof. When it comes to brain health, little lifestyle tweaks can have a profound impact. Unfortunately, there are false promises out there, filled with marketing hype, that are nothing more than a waste of money (and are sometimes dangerous), but this book dispels the myths and cuts through the noise. You'll find a map to science-supported action steps—both big and small—that can help optimize your brain function each day at any age. This integrative approach will help stave off memory problems, fight off depression, improve mood, and boost day-to-day productivity and energy. It's an inoculation against dementia and non-genetic Alzheimer's. The action steps you take to protect your brain years down the road will also optimize your brain health today and tomorrow so you can always be the best version of yourself.

Plus, there's a bonus Age-Proof Brain Seven-Day Challenge to put these recommendations into action and kick-start your brain boost.

Throughout the book, look for boxes titled "The One Important Sheet of Paper." These recommend topics you can jot down on one piece of paper, which you'll bring to your next checkup with your primary care provider. Studies suggest that tracking these critical tests and keeping them within normal ranges plays a significant role in slowing down the aging of your brain.[10] At the end of the book, you'll find that list collected in one spot. Make a copy or take a picture of it on your phone to take with you to the doctor.

This book is not an anti-aging book in the sense that it is against aging. There are aspects of aging that are wonderful. We gain knowledge, experience, confidence, perspective, and wisdom as we age. (For example, knowing what I know now, I would not have worn an all-white suit to my middle school dance.) We want to embrace these aspects of aging. In fact, a study found that those who had a positive attitude toward their age and the aging process had a 49.8 percent lower risk of dementia, even if they had a genetic risk for Alzheimer's.[11] The focus of this book is to slow down and delay for as long as possible the decline of the brain and body that often comes with aging. The goal is having a plump, clean brain and being happy, productive, and embracing life. We can get the best of both worlds: we can enjoy the natural evolution of our lives and live the best life possible.

This book is for my Bubbi Lorraine and for the Bubbi Lorraine in your life. Even though their memories were taken, we will never forget them. I'm writing this book with the hope that we can use the latest scientific breakthroughs to lessen the number of times we all go through this type of loss. This book is empowering you to not feel resigned to this fate. You can keep your brain young. You can help those you care about keep their brains young, too. *The Age-Proof Brain* will equip you with a new take on how to manage your lifestyle using cutting-edge information that can help you live a far healthier and prosperous life. It will help you feel like you can have your best days today, tomorrow, and into the future.

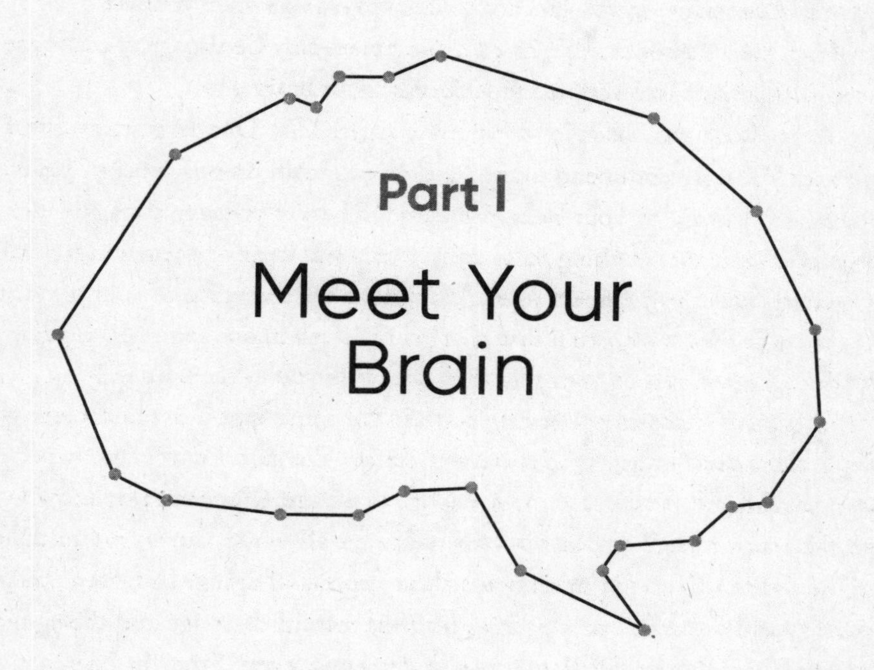

Part I

Meet Your Brain

CHAPTER 1

How Old Is Your Brain?

ROBERT MARCHAND SET A WORLD RECORD FOR CENTENARIAN cyclists when he was 101. Amazingly, even after the age of 100 he was still able to improve his speed. This centenarian would be classified as physically fit, even for a man aged 42 to 61.[1] Clearly, we can make our bodies move as if they were younger—but what about how our brains function?

And when do we even want to start thinking about our brain's age?

I often hear people in their thirties and forties comment that their brains aren't working like they used to. Maybe these people can't concentrate at work anymore or can never seem to remember if they closed the front door. Trouble focusing, diminished productivity, and difficulty with recall can be frustrating, but often these symptoms are brushed under the rug or just chalked up to stress. We all have our "human moments," but don't downplay these signs until they reach a point where they cannot be ignored. No matter what your age, your brain health today impacts your brain health tomorrow.

I'm sure you have heard the word *senility*; many people use it to describe what happens when someone has reached a certain age and their mind (especially their memory) doesn't function like it used to. You may say, "Oh, that's part of aging. There is nothing anyone can do." In fact, there is *a lot* you can do. If you take care of your brain properly, it should keep up with your body as you age. A sharp mind into old age isn't just a bit of luck—it's within reach for most of us.

For instance, in some illuminating studies, researchers performed magnetic resonance imaging (MRI) scans on participants' brains. Some of the participants then implemented certain key lifestyle changes (the same ones you will find later in this book, in fact), and follow-up brain scans taken six months to a year later showed their brains looked *younger* than before.[2] It was as if the participants in the experimental group had put their brains in a time machine.

A younger brain is plump, with more mass of gray and white matter, or brain cells. The experimental group's brains had more volume and fullness—they were actually bigger—and there was an increase in connections between their brain cells. In contrast, the participants who didn't adopt these lifestyle interventions (the control group) had visibly older brains, which had reduced in volume since the study started.

A shrinking brain loses functionality and is at risk of a long list of disorders. Think of your brain cells like tires: they need to be full to function. They're like a vibrant city bursting with activity, from building proteins to transporting molecules to replicating your DNA. It's like you have a bustling mini-Manhattan in each of your brain cells. Your ability to think, remember, and innovate is tied to the vitality and fullness of each brain cell. If the cells collapse, these functions cannot proceed and cognitive abilities suffer.

The take-home message is just because you're getting older doesn't mean your brain has to age at the same rate. Significant cognitive problems are *not* a normal consequence of aging.

Brain Development 101

To understand how your brain ages, let's take a moment and take a crash course on brain science to look at how your brain is built. I call this brain science in a couple pages, and the best part is there are no quizzes or tests.

If you were to open up your skull and look down at your brain, the first thing you might notice is that it is divided into halves separated by a long groove down the center. This entire wrinkly, pinkish-gray structure is called the cerebrum. The technical term for the halves of your brain is *cerebral hemispheres*, like the hemispheres of a globe. Just like your face has symmetry, so does your brain.

Actually, what you're seeing is called the cerebral cortex—the outer layer of the cerebrum. That's what's often called "gray matter." The inner layer is white matter, or *substantia alba*, which is made up of densely packed nerve fibers.

We can further divide the cerebral cortex into lobes. There are four key lobes on each side of your brain. Each has a different job:

- The frontal lobe, situated at the front and top of the brain, is involved in thinking, decision-making, and controlling your movement. For example, if you are trying to learn the latest dance or to play chess, you are using the frontal lobe.
- The parietal lobe, which sits behind the frontal lobe, helps you process information from your senses. If you are walking through a theme park, all the temperatures, sounds, smells, and the feelings you get when you run your bumper car into someone else are being processed by your parietal lobe.
- The temporal lobe, at the lower front of the brain, plays a role in memory. This part of the brain also takes your memories and integrates them with your senses. In chapter five, we discuss how this insight into how your brain works can help you remember like a memory champion.
- The occipital lobe is at the back of the brain. Its main role involves vision.

In the back portion of the skull, below the temporal and occipital lobes, sits the cerebellum (Latin for "little brain"). It plays an important role in movement, coordination, posture, and balance.

Finally, there's the brain stem, the lower part of the brain, which connects to the spinal cord. It regulates vital automatic processes like breathing and heart rate.

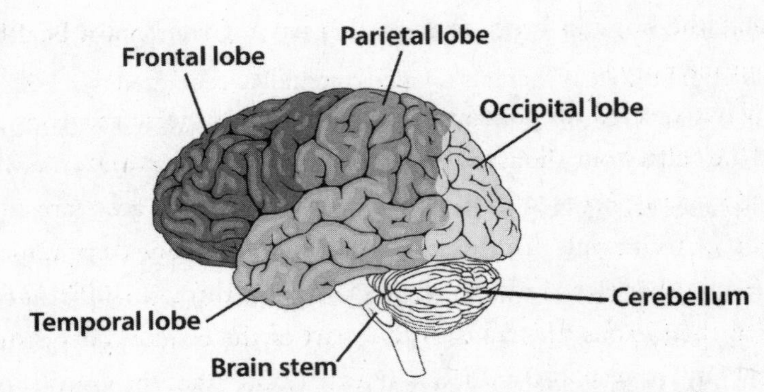

This image shows the four lobes of the cerebral cortex, as well as the lower brain structures, the brain stem, and cerebellum.

There are *many* other brain structures with other specific functions, which I'll introduce in later chapters (the brain is a remarkably complex organ). For now, you've been introduced to the basic brain parts. Let's go a little deeper and talk about what the brain is made of.

The average adult brain weighs about three pounds. Those three pounds are made up of about eighty billion brain cells, or neurons. To put the magnitude of your brain into perspective, if you could possibly count to eighty billion it would take, wait for it . . . about 2,500 years of nonstop counting. How big is one brain cell? Look at the dot of an *i* in this book. You could fit fifty brain cells in the dot of that *i*.

Here's a surprise: when you were born, your brain weighed about three-fourths of a pound.* But while your brain quadrupled in weight since then, guess what stayed the same? The number of brain cells. You have 80 to 100 billion brain cells now—and you were born with most of them.[3] That's right, you've had most of your brain cells your entire life, and they were formed before you were born. (Several parts of your brain *can* add new brain cells throughout your life, though. One of the important ones is your hippocampus, which is involved in memory. We will talk about the hippocampus

---

* Quite a bit happened before you were born. In the nine or so months of brain development you spent inside your mother, you were making a staggering 250,000 new brain cells a minute. That's a lot of work!

throughout the book in order to keep this part of your brain healthy and boost your memory.)

So, what created all that extra weight? For starters, the connections between the cells. Your thoughts, your feelings, your memories, the way you move, and every aspect of your being—essentially who you are—are all in the interaction of those cells. Those 80 to 100 billion neurons communicate by sending each other electrical and chemical signals through 100 *trillion* connections at their axons (the transmitting part of the cell) and dendrites (the receivers).[4] (By the way, 100 trillion is at least 1,000 times the number of stars in our galaxy. In fact, there is enough electricity running through your brain right now to power a light bulb. Maybe someday our brains will be powering our appliances!) As your brain developed, you also added supporting cells like astrocytes and oligodendrocytes (also known as glial cells), which hold the neurons in place. And as you learned new things, you added myelination, a coating around your brain cells that makes the electrical signals travel faster. It is similar to the insulation you see on an electrical wire. This myelination allows us to get better at tasks—*and* adds weight to the brain.*

Fast Facts About Myelination

Two things that increase myelination and improve brain processing speed? Exercise and eating fatty fish like salmon. Why fatty fish? The coating that wraps around our brain cells is omega-3, which is found in fatty fish. This is why fish is called brain food. In chapter fourteen we will delve into which foods give your brain a boost.

As the brain develops, a process called *remodeling* occurs when we learn new things. This process is most intense during our childhood and teenage years; it continues throughout the rest of our lifespan, but to a lesser degree.

---

* Humans aren't born with much myelination, which is why babies aren't very good athletes or musicians.

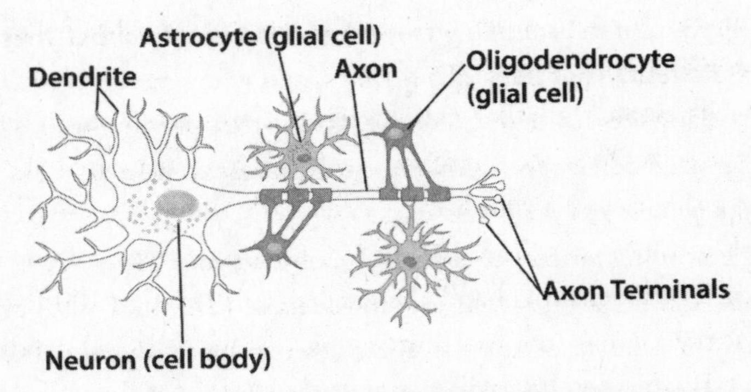

In this image of a neuron, the myelination is shown by the gray areas wrapping around the axon.

Much of that remodeling is happening in the connections between brain cells, which get stronger as you learn and remember information. These 100 trillion connections are your memories, behaviors, and thought patterns—how you move, think, and feel. How do these connections tell you where to find your keys, where you parked your car, or the name of the person you just met? In essence, how do you form and revisit a memory? Anytime you learn something new, whether it was how to walk when you were a baby or how to use that new piece of technology, you make a connection between brain cells. Each brain cell can connect to about 10,000 other cells, which is why right now you have about 100 trillion connections in your brain. But how many connections did you have when you were three?

Ten times more than you do now![5] A thousand trillion, a quadrillion. That's a number you don't hear every day. Between being born and turning three years old, connections are made at *ten to twenty thousand connections a second*. This explosion of connectivity is called *synaptic exuberance*. The process peaks somewhere around eight years old. Then, you start rapidly losing brain connections. But don't panic! This is a normal process called *pruning*, wherein the brain prioritizes which cell connections to keep and which to let go. Pruning allows your brain to both strengthen connections related to critical skills and information, and form aspects of your personality. Think of your brain like a topiary bush or bonsai tree: During these critical years of developing, your brain is clipping away wayward foliage so that your personality, your interests, your likes and dislikes, and who you are at your essence can materialize. The key take-home message is that pruning unneeded connections happens during the developmental process to help your brain shape you into the person you are becoming.

But the pruning process continues throughout your life as you make new connections and shed ones that are not needed. (Think of your brain like your desk: you need to clear away the clutter to be productive. But clutter will inevitably creep back in, so clearing it isn't one-and-done—you'll have to clear it off again . . . and again.) Every night when you go to sleep, your brain eliminates connections it figures you don't need anymore.*

As you age, your brain doesn't change as rapidly as it did when you were a child, but it still dynamically changes all the time. Because your brain is malleable, you can make it more robust and more adaptable at *any* age to improve memory, manage stress, and boost focus and productivity. You can also transform your brain at any age to adopt new habits, gain perspective, and recover from physical and mental trauma. The great hope of life is that we are always a work in progress, able to get better and be better each day as our brain remodels.

So, How Old Is Your Brain?

We know that our brains are often not the same age as our chronological age. For instance, there is a group of people called SuperAgers, who are in their eighties and beyond but have the cognitive function of those decades younger.[6] Conversely, it's possible for your brain to be *older* than your chronological age. Obviously, that's not something you want!

---

* In chapter five we delve deeper into how your brain retrieves the memories you want to access, like that famous actor's name (you know, the guy in that movie), and, of course, more serious information.

While there's no test you can take at home to definitively say how "old" or "young" your brain is,* we can think of a young and healthy brain as being at peak function. And, most often, peak brain function is associated with a sharp memory. As we get older, it can be difficult to remember names, faces, events, something we just read, or what we ate. In younger brains, the process called *neural differentiation* is efficient and robust. In this process, specific brain cells are tasked to remember certain types of information, such as faces. As we age, that process deteriorates, so the cells lose their specificity and do not function as well. Instead of just focusing on faces, they try to remember other types of information as well. For a SuperAger, neural differentiation is akin to that of a twenty-five-year-old. That's part of the reason why a Super-Ager has the memory performance of a twenty-five-year-old.

So what else are the secrets of these SuperAgers with robust memory—and all those with brain ages younger than their chronological ages? A study published in 2021 uncovered some surprising answers. Over eighteen months, the study followed 330 people, referred to as SuperAgers, who were 100 years or older; the researchers found no decline in most areas of memory or cognitive abilities.[7] (While a year and a half might not sound like a long time, once a person reaches the century mark, two years for them is like twenty-five years for a seventy-five-year-old in terms of brain health. For example, the risk for developing dementia increases by 60 percent every two years after the age of 100, while it takes twenty-five years for a 75-year-old's risk of dementia to increase by the same amount.[8] In other words, twenty-five years of risk is compressed into two years after the age of 100.)

So, what is the secret of these mentally strong centenarians? You might be tempted to guess "genes." While genes can definitely play a role, 16.8 percent of the people in the study had genes that are known to *raise* the risk of Alzheimer's, and they did not develop the disease. What seemed to be a key piece of the puzzle was lifestyle!

* Although many online quizzes might tell you otherwise—most of those just aren't based in science.

One key factor was that they kept learning new things throughout their lives. Remember, your memories are housed in those connections between your brain cells. Think of your brain like a bank account; the more deposits we make, the less our net worth is affected by withdrawals. We make deposits (new connections) by learning new things; as we age and naturally lose some of those connections, there are simply more remaining. There's a Spanish saying: "Learn one new thing each day." This simple advice is an excellent first rule for brain health. In chapter sixteen we discuss in depth which types of learning are best, but a key take-home message right now is that learning new information or a new skill keeps your brain young. So, if you are learning something new by reading this right now, you are doing one of the most important things for your brain.

While we are not at the point where every person can slide into a brain scanner and find out their brain age, here are just a couple of factors that can help give you a sense of your brain age. Ask yourself the following questions:

1. Executive functioning—How well can I manage my day?[9]
2. Balance and coordination—How well can I move and maintain balance?[10]
3. Ability to learn and recall—How well can I remember important information?[11]
4. Movement—How fast can I walk?[12]
5. Identity—How old do I feel?[13]

Of course, none of these questions can replace an actual brain scan and comprehensive evaluation by a neurologist. Still, these fundamental categories can help us get a sense of brain age. It is important to note that simply determining someone's brain age by brain scan is also not without complexity and controversy.[14] For example, there is the issue of which part of the brain and which biomarkers, such as white or gray matter, iron deposits, and volume, to investigate or prioritize. Be wary of a private clinic offering to tell you your brain age, especially for a hefty fee, as at this point this tool is primarily being used accurately at large research institutions. For that matter, be wary if it's an offer from a guy in a van for a bargain fee. In all seriousness, determining brain age using cutting-edge imaging technology has become an emerging and powerful tool in research studies at large institutions to understand health, disease, and mortality.[15]

We will continue to discuss other key factors that provide insight into our brain age. Remember, it is not just one of these factors but a combination that determines brain age.

Brain Trash and Brain Age

Now that you know your brain can be older than your chronological age, you may be wondering how that happens. One of the things that makes a brain older, or age prematurely, is the buildup of trash and toxins (we discuss other factors in the coming chapters of this book). The trash is a by-product of the work your brain cells do. Remember, we said each of your brain cells is like a hustling, bustling Manhattan. Just like how a city can get dirty, in the process of living your brain gets dirty as well—filled with waste in the form of leftover chemical reactions, environmental toxins, old, damaged cells, and no-longer-needed proteins. Your three-pound brain makes five pounds of trash a year. This trash is normally recycled or flushed out, but if those processes break down, the trash can build up and damage the brain. The trash buildup interferes with your brain cells' ability to communicate with each other, eventually causing the cells to shrink and die. There are many forms of brain trash, but the two key types are amyloid plaques and tau tangles.

Plaques

To better understand where plaques come from, imagine a house with an antenna on the roof that delivers some basic television TV stations. If this antenna breaks, not only will you not be able to receive the TV stations, but broken pieces from the antenna could damage your roof. The same idea can happen with your cells. The surfaces of your cells have receptors that look like antennae, which function as an extensive communication and security system. There are many different types of receptors that play a role in all the

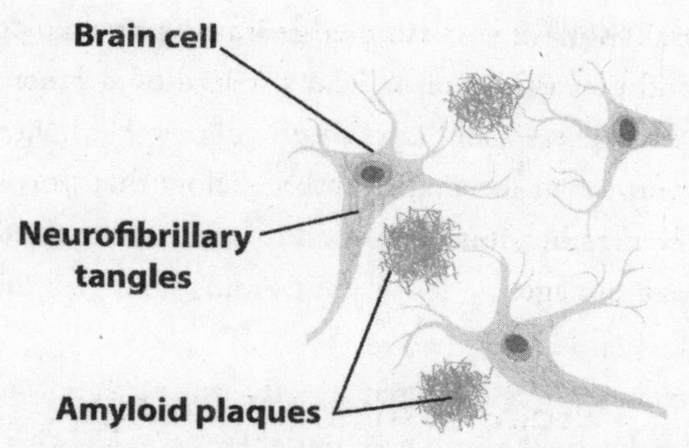

These plaques turn into garbage that gets in the way of the brain cells' ability to form the connections that make and remember memories.

complex actions and emotions of life. For instance, some receptors specifically receive signals that tell the cells whether or not to grow. Receptors also attach to chemicals trying to enter the cell. If the receptor recognizes the chemical, it unlocks the cell so the chemical—which can impact functions ranging from mood to learning emotions such as love*—can enter.

One type of receptor found on brain cells is called amyloid precursor protein, or APP. Sometimes enzymes, acting like scissors, come along and chop up APP into little pieces. There are several factors that cause these enzymes to chop up the APP, one of which—inflammation—we discuss in the next chapter. What's important to know now is that the chopped pieces of this specific receptor APP happen to be sticky and clump together to form plaques that we call beta-amyloid.

Tangles

While plaques form between brain cells, tangles are trash that forms *inside* the cells. Inside your cells you have an extensive system of filaments and tiny tubes, called a cytoskeleton, which delivers proteins and nutrients to different

* When people say love is mysterious, they are really saying love is specific chemicals released in the brain that attach to receptors. Not so romantic, right?

parts of the cell. Think of this cytoskeleton as subway tracks, but a microscopic version. The cytoskeleton is held in place by a protein called tau—which in this analogy is much like the railroad ties that hold tracks in place. Sometimes the tau molecules that hold this elaborate delivery system in place detach from the cytoskeleton and attach to other tau molecules. Imagine a bunch of railroad ties coming loose and forming a tangled mess; that's what a "tangle" in the brain looks like.

It is not completely clear what exactly causes these tau molecules to no longer hold the tracks in place, but there is evidence that inflammation, buildup of toxic components, and chemical reactions in cells are all factors. Once tangles accumulate in one brain cell, they then leak out of the cell and get taken up by a neighboring healthy cell. This healthy cell gets tricked into making more tangles and thus the damaged cells spread throughout the brain.[16] A 2021 study found that tracking the accumulation of tau protein in the brain is a better predictor of memory decline than tracking plaque formation.[17]

Plaques and tangles are the two most common forms of brain trash found in Alzheimer's disease (we discuss this more in chapter six), and tau tangles are also found in other forms of dementia, as well as chronic traumatic encephalopathy (CTE—a condition where brain damage occurs due to trauma from head injuries such as concussions). But there are also other types of brain trash that can be a root cause of a wide range of diseases and dysfunctions, including depression, Parkinson's disease, and anxiety, to name just a few. For example, a protein called alpha-synuclein accumulates in toxic clumps that damage dopamine-producing brain cells.[18] Dopamine helps coordinate movement. When these cells are destroyed, the tremors seen in Parkinson's appear.

Buildup of waste in general can also lead to symptoms such as diminished focus, loss of productivity, and a drastic reduction in overall energy—signs and symptoms of biological aging. Simply put, the more trash, the "older" the brain.

So, can we get rid of trash? While there's no brain shampoo, scientists have made a series of discoveries that show your body has powerful cleaning methods to literally take out the trash. Learning new things, which also increases the connections between your brain cells, is one of them! The next natural cleaning method is related to your immune system—which we turn to in the next chapter.

Plot Twist!

I just said that more brain trash means an older brain . . . but before we move on to chapter two, I have to acknowledge that that's not *always* how it works (and here's where the discussion of plaques and tangles starts to get somewhat perplexing—bear with me). Let's go back to the study of the SuperAgers, who have the memory recall ability of someone much younger. Some of them had enough plaques and tangles in their brains that they should have been guaranteed to have memory dysfunction. Yet they didn't. It seems that these SuperAgers have brains that can be resilient to higher levels of trash—trash is present but doesn't seem to be interfering with memory. Additional research from other studies has found that 30 percent of older adults have enough plaques and tangles in their brains to have significant memory loss but don't show signs of dementia.[19] Again, this resilience can be attributed to their lifestyles and possibly some aspects of genetics. The lifestyle factors that can compensate for, and build resilience to, plaques and tangles are covered throughout this book.

To be clear, we definitely don't want too much trash to build up. It can damage the brain, though certain types of waste and trash are more dangerous than others. But we are learning that trash is not the *only* factor in brain aging and brain dysfunction—we cover others in the coming chapters.

In part three, we focus on how to get rid of trash and how to keep your brain youthful so that you're more resilient to the trash you *can't* get rid of.

CHAPTER 2

Your Brain, Your Immune System, and You

PONCE DE LEÓN SCOURED THE WORLD LOOKING FOR THE FOUNtain of youth, but it turns out that he should have looked closer to home. In fact, Ponce could have stayed on his couch and just studied his own immune system. Breakthrough studies have discovered that the fountain of youth is actually a balanced and healthy immune system.[1] Just as your brain can be younger or older than your chronological age, so can your immune system. One year from now, you may have lived 365 (or 366!) more days, but that doesn't have to mean your brain, body, and immune system aged a year. A study published in *Nature Genetics* found certain remarkable centenarians with immune systems several decades younger than their actual age—on average, forty years younger, to be precise.[2] One very healthy 105-year-old man even had the immune system of a 25-year-old.

It's important for brain health to have a young immune system because, as the immune system ages, it either becomes over- or underactive. This unbalanced immune response raises the risk for infection, dementia, Alzheimer's, Parkinson's, heart disease, depression, and cancer, among others. Let's look deeper into how the immune system works—and how it protects the brain—so that we can keep the immune system balanced and young.

A Crash Course on Your Immune System

Your immune system is a complex army with a dizzying number of players. The main parts of the immune system are:

1. White blood cells: These amazing cells battle infection and disease. There are many types of white blood cells, including neutrophils, eosinophils, monocytes, T cells, and B cells.[3] You don't need to remember all those names—just that it's a complex army.
2. Antibodies: Made by white blood cells, antibodies are proteins released into your blood. These proteins recognize and destroy foreign substances such as bacteria and viruses.
3. Red blood cells: It's now known that red blood cells do more than carry oxygen and assist in clotting. These cells also play a role in the immune system by attaching to dangerous pathogens and aiding in their removal.[4]
4. Spleen: The spleen helps control the number of white and red blood cells in your body and removes old or damaged blood cells. It houses antibody-producing lymphocytes, among many other functions.
5. Thymus: A gland in your chest that makes white blood cells. Hormones released by the thymus also inhibit the aging process.
6. Lymphatic system: This large network of organs, lymph nodes, and tissues functions like the surveillance and sewer system in the body. The lymphatic system is part of the immune system. Although that comparison might sound a little gross, this system plays a particularly noteworthy role in our brain health—as we will see in the next section.

In this chapter, we cover just a few components of the immune system that have particularly noteworthy roles in our brain health. In addition to the lymphatic system mentioned above, we also look at some of the specific cells that participate in our body's immune response.

The Lymphatic System: Taking Out the Trash

Each day, the glands, tissues, and organs that make up the lymphatic system clear the waste and toxins that build up in our bodies. Think of the lymphatic system as plumbing pipes with lymph running through them. Lymph is not blood—it's a colorless fluid that is drained from your tissues and organs. Lymph removes the by-products generated by your cells, as well as toxins and germs. It circulates through the body and picks up dangerous viruses and bacteria, depositing them in the lymph nodes (which are located in the neck, armpits, chest, abdomen, and groin). The immune cells sift through the viruses and bacteria in the lymph nodes; if the immune cells find anything dangerous, they mount an attack on these pathogens still present in the rest of the body. I like to think of these immune cells as miners panning for gold, but instead of looking for nuggets, they are searching for pathogens.

This process is why you may get swollen nodes in your neck when you have a cold: blood cells are rushing into the lymph nodes to fight off the infection, and the pileup causes swelling. It's a sign your body is working to get rid of the germs that are making you sick. The lymphatic system also cleans up the mess left behind after injuries to the brain and body. And, importantly, part of the waste that the lymphatic system removes is the brain trash I described in chapter one.

For hundreds of years, however, scientists believed the lymphatic system did not extend into the brain.[5] As I mentioned in the introduction, scientists and doctors have had a tendency to view the body as a series of silos, where one system, organ, or process works in isolation from all the others. That's exactly how experts *had* viewed the lymphatic system until 2013, when Dr. Maiken Nedergaard at the University of Rochester made a startling discovery. Dr. Nedergaard was trying to understand how the brain repairs itself after an injury. She wanted to investigate how sleep played a role in the repair process. While in deep sleep, her subjects experienced a phenomenon that seemed like something out of a horror movie.

She observed that the brains of the sleeping subjects shrank down to 65 percent of their normal size. Once the brain reduced in size, it began to pulsate in a dynamic rhythm. (I like to imagine this happened on a stormy night, with flashes of lightning and thunder echoing through the lab. But likely it wasn't that dramatic of a scene.)

Further research discovered that the shrinking process squeezed out the trash, waste, and toxins from the brain cells and created empty space into which the garbage could drain. Once the debris was squeezed into these empty portals in the brain, while the volume of the brain was still reduced, fluid from the spinal cord plunged into the brain for a brain wash.

The idea that our brains squeeze out waste and are flushed clean every night went against everything we thought we knew about how the brain works. In fact, the scientists who made these discoveries didn't believe their results and spent several years repeating their experiments because they figured something had to be wrong.[6]

The answer to how and why our brains shrink when we sleep arrived in the form of a flashing green light. In 2015, Dr. Kari Alitalo set out to take a closer look at the lymphatic system in his laboratory in Helsinki, Finland. Using cutting-edge technology, he was able to map the lymphatic system of mice, turning all of the tunnel-like vesicles of their lymphatic systems green and creating a detailed map. Dr. Alitalo expected to see the lymphatic system light up from the neck down. To his surprise, the glowing green light also showed up in their brains.[7]

We now call the part of the lymphatic system that extends into the brain the glymphatic system, as it works with glial cells in the brain. Glial cells are supporting cells that have a range of functions such as protection, cleaning up debris, and forming myelin. The discovery of the glymphatic system set in motion a dramatic change in our understanding of how our

Where Does All This Trash Go?

In 2015, scientists discovered a set of vessels that run from the brain into the lymphatic system. They are called meningeal lymphatic vessels and, through them, cerebral spinal fluid drains from the brain, removing the waste.[8] Think of these vessels like the pipes attached to a dishwasher that remove the waste water after you run the wash. The age of our brain and body is related to the age of these vessels. Think back to the pipe analogy: If that pipe doesn't work, the sewage builds up. Imagine the brain not being able to remove waste and how that could cause damage and aging.

Just as an aside, it's incredible that these vessels were only discovered so recently. We thought all of human anatomy had been completely mapped, but these vessels, hidden deep within the brain by other blood vessels, were tough to locate.

bodies work and how we can lower our risk of a range of neurological diseases. Anything that disturbs our waste removal system—from aging, to concussions and brain injuries, to a stroke (which can lead to what can be described as the brain drowning in fluid)—is bad for our brain health.[9] In short, because our lymphatic system plays such a big role in clearing out brain trash, a malfunctioning "sewer" system can lead to a variety of brain dysfunctions.

Furthermore, the link between the immune system and the brain can help us better treat and understand a range of seemingly disconnected conditions, from dementia to depression to multiple sclerosis to concussions. An immune system that incorrectly attacks the brain does damage. And an immune system that does not remove waste and toxins also causes damage. So, to effectively treat these conditions, a delicate *balance* of protection and cleanliness must be maintained in the brain.

Microglia: The Cleanup Crew

Besides sleep, there is another way to remove waste and garbage from your brain. Have you ever looked in an aquarium and noticed those little fish at the bottom of the tank that are eating up waste and garbage? You have cells in your brain called microglia that perform a similar function. These microglia are immune cells that basically act as scavengers in the central nervous system, eating up brain trash.

However, they can get confused, and instead of eating trash, plaques, tangles, and toxins, these cells start to gobble up healthy brain cells and turn them into more useless brain trash. We discuss in chapter eight how to keep your microglia from getting confused.

Killers and Peacekeepers

The next key players in our immune and brain health are immune cells. There are two key categories that maintain a balanced immune system: pro-inflammatory (killer) cells and anti-inflammatory (peacekeeper) cells.

One of the critical types of killer cells are the T cells, which circulate in the body to seek out and destroy specific viruses and bacteria they have learned to recognize. T cells rush to the site of an infection to trap and destroy the invaders, a battle that generates inflammation, which can physically manifest as pain, swelling, and redness. So, when you get a sore throat, it's not actually the infection causing all of the soreness but rather killer cells attacking the virus.

The attack itself kills the infected cells, as well as some neighboring cells, which are replaced with healthy, uninfected cells. Your body senses that the virus is being or has been eliminated. Then, it's time for peacekeepers to do their job. These anti-inflammatory cells rush to the scene and calm down the killers, basically telling them that they have done a good job killing—and it's time to go home, kick back, and relax.

When people talk about boosting your immune system, they're really talking about boosting those killer cells. However, if the peacekeeper cells are not deployed, your throat would be continually attacked and damaged by your own immune system. When this kind of continual attack happens, it's called chronic inflammation. There can be too many killer pro-inflammatory cells or not enough peacekeeper anti-inflammatory cells to keep inflammation under control throughout the body. That's why it's necessary to *balance* the pro- and anti-inflammatory responses.

Get Enough Vitamin D

There is a lot of marketing and exaggeration in the wellness world, but Vitamin D isn't just hype—it really does help support immune health. Vitamin D is essential to the proper functioning of your T cells. Imagine your T cells like little cars driving around inside your body looking for dangerous viruses and bacteria. T cells each have their own antennae-like receptors, which reach out and grab vitamin D from the blood. The vitamin D fuels T cells to keep them moving and focused; if there's not enough vitamin D, the killer T cells can't do their job attacking dangerous viruses and bacteria.[10]

Vitamin D is sometimes called the sunshine vitamin because our bodies manufacture 50 to 90 percent of their vitamin D in response to the sun hitting our skin. Less sun exposure can lead to our bodies manufacturing less vitamin D. The lack of sunshine exposure is one reason we get more colds and flu in the winter.[11] We discuss other key vitamins for brain and immune health in chapter fourteen.

Problems of an Aging Immune System

There is another key player in the immune system we need to discuss. (I know what you're thinking: *Another player? This feels like the beginning of the movie* Encanto, *when twenty characters are introduced in two minutes.* While

Lin-Manuel Miranda does an amazing job of clarifying all of these characters in a fantastic opening number, unfortunately—though fortunately for your ears—I can't sing the immune system to you. But stick with me, this is the last new player for this chapter!) This new player has an important role that's akin to a security guard on duty. If the guard notices something wrong or some suspicious activity, they call in backup. The security guards of your immune system are called cytokines. These players alert the killers, such as T cells, that it's time to destroy an invading germ.

But, as you know by now, balance is the key. If cytokines overreact to an infection, they can trigger an overly large immune response. Imagine you called 911 because you had a break-in at your house and instead of sending a couple of police officers, the 911 operator sent an entire platoon. The army would do more damage to your home—and possibly to you as they trampled around—than the burglars. Similarly, an excessive immune response can damage organs such as the brain.

A significant reason our immune system miscalculates is the aging process.[12] If the glymphatic system fails, we are not able to adequately remove trash and toxins from the brain. An older immune system is also more likely to overreact or underreact to viruses and bacteria, which can have serious consequences: a weak immune system can allow viruses, bacteria, and cancer cells to grow and replicate, making us sick. A too strong or overactive immune system can attack our joints, organs, or brain, resulting in inflammatory and autoimmune diseases. (Autoimmune diseases such as rheumatoid arthritis, lupus, and inflammatory bowel disease, in all their complexity, are essentially those killer cells attacking one's own body.) Obesity, heart disease, and diabetes can likewise be caused or partially caused by the immune system being out of control and attacking specific organs. In many cases, a factor in Alzheimer's disease is an unbalanced immune system that attacks the brain.

Depression can also be caused or exacerbated by an overactive immune system. A 2021 study found that when the immune system incorrectly attacks parts of the brain involved in mood, the symptoms of depression can manifest.[13] This is why individuals with depression sometimes do not respond to commonly prescribed antidepressant medications and talk therapy—neither of those methods address an underlying immune issue. These insights show how intimately the brain and immune system are connected—and, more importantly, provide hope in developing more effective treatments for depression.

Unsurprisingly, there are links between the immune system and memory: In a 2019 study, researchers blocked overactive immune systems in elderly mice. These mice started remembering how to navigate a maze like their youthful counterparts.[14] In other studies, mice that had a compromised immune system* scored lower on memory tasks than other mice. When researchers transplanted immune cells into these mice, essentially giving them functioning immune systems, they scored as high on memory tasks as healthy mice.[15]

In short, if you manage the immune system, you de-age the brain. It's a powerful reminder that—as I've said before—our body systems are intricately interconnected, and what happens in one area will impact another. In part three of this book, I give you an action plan on how to balance your immune system, but first let's keep following these threads that connect our brains to other body systems, shall we? Up next, we'll look at the intriguing connections between the brain and the heart.

---

* In case you were wondering, mice and humans are remarkably genetically similar—we share over 90 percent of the same genes. So, scientists have long used them in studies of the immune system, which have led to big breakthroughs in our understanding of immunity. For this research, immunologists specifically bred mice with compromised or abnormal immune systems. These mice must be kept in a special germ-free environment.

CHAPTER 3
The Heart-Brain Connection

A MEMOIR CALLED *A CHANGE OF HEART* RECOUNTS THE FASCINATING experience of Claire Sylvia, a dancer who received a heart transplant. After the transplant, Claire mysteriously started craving fried chicken and beer, which had never been her cup of tea. Perplexed by this strange turn of events, Claire sought out the family of the person she'd received the heart transplant from . . . only to find out these foods had been her donor's favorites. (You may be raising an eyebrow reading this. But Claire's story is far from the only one like it: neuropsychologist Paul Pearsall wrote a book about multiple examples of heart transplant recipients taking on traits of the donors, including mannerisms, food preferences, and interests.[1])

These anecdotal stories are not meant to provide concrete insights (unlike scientific studies). As thought-provoking as these connections are, and as wonderful as it is to contemplate mysterious aspects of the body that we don't understand, they definitely require rigorous research and analysis before we can draw any strong conclusions. But here are some intriguing facts: The heart contains about forty thousand neurons called sensory neurites, which relay information to the brain. Some researchers refer to them as the "little brain" on the heart.[2] Further research is needed to elucidate the scope of the information relayed, but one thing we profoundly and clearly understand is that heart health is one of the most powerful aspects of brain health.

In fact, we often treat our heart like it's a brain. We say things like "Follow your heart." We equate the heart with emotions, especially feelings of love.

These might sound like mere analogies, but as it turns out, the notions have some real biological underpinnings. The heart and the brain are literally connected via the vagus nerve, and they're in constant communication. Think of your vagus nerve as a telephone line that allows the brain and heart to talk to each other. For example, physical exertion such as a sprint causes a rapid heart rate, which in turn sends a signal to your brain to slow down. The heart also responds to your mental and emotional experiences. Imagine yourself about to give a big speech and how you might feel. If you are nervous or excited, your brain sends signals to your heart to beat faster.

This connection is why being "young at heart" can help you be "young at brain." What happens in your heart can impact your risk for a wide variety of brain health conditions, from dementia to depression to anxiety. Simply put, without a healthy heart it is virtually impossible to have a healthy, youthful brain.

Between the Heart and Brain, There's a Two-Way Street

The vagus nerve* isn't the only connection between the heart and the brain. Your circulatory—or cardiovascular—system is 66,000 miles long. The Earth's circumference is about 25,000 miles. That means your circulatory system could wrap around the globe more than two times. With each beat

* The name *vagus* comes from the Latin term for "wandering," and it's an apt moniker. The vagus nerve is the longest of the cranial nerves and "wanders" from the back of your brain to your heart all the way down to your gut. We talk more about the vagus nerve in chapter four, The Gut-Brain Connection.

of your heart, you are pumping nutrients, oxygen, carbon dioxide, and hormones throughout those 66,000 miles of vessels, veins, and arteries. A critical job of this system is to connect your heart and brain. While your brain is only three pounds, a small fraction of your overall body weight, it consumes about 20 percent of the oxygen made with each beat of your heart. And when you are thinking really hard, you send even more oxygen to your brain because the energy your brain needs to run comes from oxygen. Brain cells only have a very small reserve of oxygen, and they need to be constantly replenished in order to communicate, create memories, think, and do all the other amazing things your brain does. Without your heart pumping oxygen to your brain, your brain wouldn't be able to survive.

Heart disease is the number one cause of death worldwide. In the United States, every thirty-six seconds someone dies of cardiovascular disease.[3] Furthermore, heart disease is one of the key driving forces of brain dysfunction. The evidence for the heart-brain health connection is overwhelming. Coronary heart disease is associated with a 40 percent increased risk of dementia, and heart failure increases the risk for dementia nearly twofold.[4] Reduced blood flow to the brain is correlated with a buildup of tau tangles, the waste that forms inside brain cells.[5] Furthermore, a stroke can increase the production of plaques in the brain associated with Alzheimer's.[6] Similarly, untreated cardiovascular disease, high blood pressure, atrial fibrillation (A fib), and smoking leave the brain starved for oxygen and are significant risk factors for Alzheimer's and dementia, depression, anxiety, and brain dysfunction.[7] If the heart and blood vessels are damaged, it's very likely the vessels of the brain are damaged as well.

We see the importance of the heart-brain connection in many surprising ways. Baby boomers scored *lower* on their memory tests than any generation in the past 150 years.[8] A critical reason for this *memory* decline is the impact of prevalent *heart* disease. If the brain cells aren't continually receiving oxygen, they can't make the new connections where new memories are stored. Furthermore, this study followed participants for six years and found that people with healthy hearts performed significantly better on memory and cognitive tests throughout the study. In contrast, those with low-quality heart health had a significant decline in memory and executive functioning—the ability to run one's day through planning, focusing, remembering instructions, and multitasking.[9]

The relationship between the heart and the brain is a two-way street. For example, depression and anxiety worsen the prognosis of heart disease, while heart disease raises the risk of depression and anxiety.[10] Likewise, while it's no surprise that when you are nervous your heart rate goes up, having an abnormally elevated heart rate can send signals to the brain that either create or magnify the feeling of anxiety. Controlling our breathing through calming breathing exercises reduces elevated heart rate and sends calming signals to the brain. We will discuss in depth in chapter nine how depression and anxiety, when not effectively treated, can age the brain.

If you think back to chapter one and the studies with centenarians who had quite a number of plaques and tangles in their brains but no significant memory loss, you'll remember that plaques and tangles alone are not a direct

The Heart-Immune Connection

Unsurprisingly, given the connections between both the immune system and the brain and the heart and the brain, the immune system also plays a crucial role in heart health. Inflammation caused by an overactive immune system can damage arteries and blood vessels in the heart and the brain. A study found that people with autoimmune diseases were 53 percent more likely to be admitted to a hospital due to cardiovascular disease and were 29 percent more likely to develop vascular dementia caused by poor blood flow to the brain than people without autoimmune diseases.[12] It is important to note that in the study not all autoimmune diseases were linked to an increase in dementia. For example, those with rheumatoid arthritis had a 10 percent lower risk of developing dementia. The authors of the study theorize that the medications used to treat rheumatoid arthritis lower inflammation and thus might lower risk of dementia.[13]

route to dementia. In fact, even with an accumulation of brain trash, your risk of developing Alzheimer's is lower if you have a healthy heart and higher if you have cardiovascular disease. This is one of the compensatory mechanisms where a healthy heart can make the brain resilient to brain trash. Researchers analyzed 395 studies and identified twenty-one factors that mitigate Alzheimer's and dementia risk. *Two-thirds* of the preventative measures were related to heart health.[11]

Seven Risk Factors for Heart and Brain Issues

Since your heart and brain health are intertwined, how do we keep your circulatory system—your heart, blood, and blood vessels—as healthy as possible? These are the seven areas where protecting your heart means protecting your brain:

1. Cholesterol
2. Blood pressure
3. Heart rate
4. Blood sugar
5. Homocysteine
6. Smoking
7. Weight

1. Cholesterol

When we hear the word *cholesterol*, we often associate it with something that's unhealthy, like fried food. But cholesterol is critical to the health of your brain and nervous system. The myelin we discussed in chapter one, which coats your brain cells, is made of cholesterol. Also, the neurotransmitters dopamine and serotonin use cholesterol to communicate. So we do need some! It's all about balance. Essentially, there are two types of cholesterol. There's the good (HDL) and the bad (LDL). A 2021 study analyzed almost two decades of data and found no association between HDL levels and dementia, but found a significant association between high LDL levels and dementia.[14] It's essential to keep track of both HDL and LDL numbers through blood tests to try to keep these numbers in the normal range.

LDL (low-density lipoprotein) cholesterol builds up in the arteries and forms plaque, which blocks blood flow to the brain. Oxidation is what causes a pipe to rust; accumulation of LDL is oxidation to the pipes of your arteries. You definitely want to be on top of your LDL levels, especially as evidence suggests that high LDL levels are also associated with an increase in the amount of plaque formation in the brain itself.[15]

HDL (high-density lipoprotein) is the other type of cholesterol. I like to think of it as the Uber or Lyft of cholesterol. It swims around in your bloodstream, picks up the LDL, and takes it to the liver to be processed. We want higher HDL levels so that it will pick up the bad cholesterol before it attaches to the arteries. It can be tough to remember the difference between the two types, so one helpful trick is to focus on the first letter: we want to keep *L*DL *lower* and *H*DL *higher*.

In part three, we discuss ways to keep your cholesterol levels in normal ranges, but here's a quick tip from two different studies published in 2021: Nuts! One study found that eating a half cup of walnuts daily for two years lowered LDL.[16] The other study found that those who were at risk of cardiovascular disease and ate pecans for just eight weeks showed a significant reduction in LDL.[17]

What About Statins?

Statins are among the most common medications used for treating high cholesterol, yet fewer than half the people who get this prescription fill it. Why? Part of the reason is the alarming headlines that statins harm memory or raise the risk of dementia. A comprehensive analysis published in the *Journal of the American College of Cardiology* looked at over a thousand elderly individuals. For over six years, the researchers measured five areas of memory using thirteen different methods, including brain scans and memory tests. The study found no link between statins and memory loss.[18]

Your One Important Sheet of Paper: Test Cholesterol

HDL and LDL are critical to check; make sure to put them on the One Important Sheet of Paper that you bring with you to regular doctor visits. Your health-care practitioner can check these with a blood test called a lipid profile or lipid panel. Interpreting your results may be a bit complicated—the optimal levels of HDL and LDL vary from person to person, depending on your sex and whether you have coronary artery disease (including a past heart attack) or diabetes. Your health-care team will help you understand your results.

In fact, if statins are used appropriately, there is evidence they are *protective* against memory decline and dementia in some individuals. If you are taking statins and notice any changes to memory or cognition, report them immediately to the physician who prescribed it.

An important caveat: a study published in *Diabetes/Metabolism Research and Reviews* found long-term use of statins can raise the risk of elevated blood sugar levels and type 2 diabetes.[19] The study's authors stated that they would never suggest that people stop taking statins, as their data show that the benefits can outweigh the risks. However, if you are taking statins, have your blood sugar monitored. (See "Your One Important Sheet of Paper: Testing Blood Sugar" on page 87.)

2. Blood Pressure

Whether it's too low or too high, blood pressure is a risk factor for brain dysfunction. Blood delivers oxygen to the brain, and you need to send enough to keep your oxygen levels heathy. When you stand up, do you consistently have a light-headed or dizzy feeling? If so, you may have orthostatic hypotension caused by low blood pressure. When blood pressure drops, not enough blood is supplied to the brain.

Conversely, high blood pressure damages blood vessels in the brain (thereby shrinking the brain) and can cause white spots, which may indicate blood vessel disease. Approximately 50 percent of the population has high blood pressure—and many of these individuals are not aware their blood pressure is high. But blood pressure is something that we can control, and controlling blood pressure in early and midlife can protect the brain years later. Lower blood pressure can reduce the risk of developing dementia by 7 percent over four years.[20] A study looked at people in their thirties and found a blood pressure reading of 110/70 correlated with a younger-looking brain as compared to a reading of 135/85.[21] Another study looked at a group of individuals aged eighteen to thirty and followed them for thirty years. Those who had high blood pressure in early life had premature aging of their brain in midlife.[22]

A Quick Tip on Checking Your Blood Pressure

When doing a home or blood pressure reading at the pharmacy, check each arm. Sometimes a reading can be inaccurate, and studies have shown that checking both arms gives you a much more accurate reading. If you get two completely different readings, repeat a third time to make sure you receive a reading that is the same at least twice.

Is It High Blood Pressure . . . or Is It Something Else?

Almost 50 percent of women develop high blood pressure before they turn sixty, but the condition may be missed or misdiagnosed.[23] Symptoms such as hot flashes and palpitations are often thought to be related to menopause rather than being identified as symptoms of high blood pressure or hypertension. High blood pressure is one of the greatest mortality risk factors for women.[24] High blood pressure can begin around menopause, but if it is not treated effectively, it can raise the risk of brain dysfunction, cardiovascular disease, and death. We have excellent lifestyle and drug therapies for high blood pressure—so long as it's diagnosed and treated in time.

For example, using a breathing device for just five minutes a day has been shown to lower blood pressure as effectively as aerobic exercise or medication.[25] This intervention is called high-resistance inspiratory muscle strength training (IMST), and it works by strengthening breathing muscles, which in turn lowers blood pressure. One study found that postmenopausal women who are not taking supplemental estrogen do not benefit from aerobic exercise programs the same way that men do in terms of heart health.[26] The study specifically looked at vascular endothelial function, which is the health and function of arteries. However, researchers have since found that using the IMST device improved vascular endothelial function as much in women as it did in men. After using the device for five minutes a day for six weeks, participants averaged a 45 percent improvement in vascular endothelial function, while they also showed lower inflammation and improvement in brain function and physical fitness. Altogether, this points to the potential of the device to help aging adults ward off cardiovascular disease. Further studies are being done, but it is worth discussing this device with your physician.

3. Heart rate

Your heart beats about 115,000 times a day, and with every beat, it sends about 20 percent of the oxygen in your body to your brain. The *only* way to supply your brain with the oxygen it needs is to have a healthy heart and blood vessels.

A study published in 2021 in the journal *Alzheimer's & Dementia* followed older adults for twelve years and found that an elevated heart rate while resting of 80 beats per minute or higher was correlated with a 55 percent higher risk of developing dementia, when compared against those with a resting heart rate of 60-69 beats per minute.[27] (Although traditionally a "normal" resting heart rate is considered 60 to 100 beats per minute, this study suggests that for older adults, the target range is lower, 60 to 80 beats per minute.) Since heart rate is easy to check and lower, this is a route to improve heart and brain health.

One risk factor for a high heart rate is atrial fibrillation (A-fib). A-fib is caused by chaotic electrical signals that lead to a fast and irregular rhythm, with a heart rate that ranges from 100 to 175 beats a minute.[28] It is a risk factor for dementia and cognitive decline, as it can change the amount of blood flow to the brain.

An elevated heart rate, as we've touched on earlier, increases the activity of the nervous system, which can raise the risk of anxiety and mental health disorders. Remember: the brain is reacting to the heart rate. An elevated heart rate is like an alarm that is telling the brain to worry—and an elevated heart rate and blood pressure increase the risk for anxiety disorders, obsessive-compulsive disorder, and schizophrenia down the line.[29]

4. Blood Sugar (Blood Glucose)

Sugar, or glucose, fuels every cell in your body, and, as you can guess, your brain cells use the most sugar. Despite making up just a tiny fraction of your overall body weight, your brain uses half of the sugar that enters your body.

Brain cells use sugar to make neurotransmitters, which allow you to think, remember, and focus. Low blood sugar, or hypoglycemia, makes it difficult to concentrate and pay attention. If blood sugar drops too low for too long, damage can be done to the brain.

On the other hand, high levels of sugar in the blood damage arterial walls and can lead to inflammation in the arteries. This inflammation also weakens the arteries, so they cannot deliver enough blood to the brain. In chapter fourteen, we delve into the sugar question in more detail and cover take-home tips to keep your blood sugar at healthy levels. For now, think of sugar like the gasoline you put in your car: it's the fuel for your brain, heart, and the rest of your body. You need to fill up the tank—but not to the point of overflowing.

5. Homocysteine

Your body needs an amino acid called homocysteine to produce proteins, the building blocks of life. Your body produces homocysteine, but you can also get it from meat. Normally, your body quickly breaks down homocysteine into other substances your body needs, so there's not much of it in the blood—or at least, there shouldn't be. One of the most important and widely overlooked avoidable risk factors for dementia is an elevated homocysteine level in the blood: high levels contribute to blood clots and blood vessel damage.

If you have high homocysteine levels, your treatment plan will depend on the underlying cause. One of the most common causes is vitamin deficiency—your body needs vitamins B6, B12, and folate to break down homocysteine. If high homocysteine levels are associated with a vitamin deficiency, there's a simple way to lower them: include enough B complex vitamins in your diet. B vitamins help your body obtain energy from the food you eat, and they also make red blood cells.

Your One Important Sheet of Paper: Test Homocysteine

A simple blood test can determine homocysteine levels; add it to your list of tests to ask your doctor about at your next physical. Healthy homocysteine levels are usually between 5 and 15 micromoles per liter; men tend to have higher levels than women, and levels may go up naturally as we age. Your doctor can work with you to identify the cause of elevated homocysteine and a plan to bring them down.

Foods High in B Complex Vitamins

- Sweet potatoes
- Almonds
- Peanuts
- Tuna
- Sardines
- Eggs
- Leafy green vegetables
- Milk
- Oysters, clams, mussels
- Legumes

B vitamins can be increased either through food or supplements. To improve absorption of B vitamins, it is imperative to include sufficient omega 3 fats, which are contained in foods such as avocado, salmon, walnuts, and olive oil (we discuss healthy fats in chapter fourteen.)

6. Smoking

It's likely not a big surprise that smoking raises the risk of heart disease and increases the risk of brain dysfunction. While smoking is often branded as something that helps a person relax by easing tension, it actually interferes with brain chemicals that balance mood. Thus, smoking increases feelings of stress and anxiety, also putting smokers at a higher risk of developing depression; a 30 percent higher risk of developing dementia; and a 40 percent higher risk of developing Alzheimer's disease than nonsmokers.[30] Smokers also put those around them at risk: the estimated billion smokers in the world create secondhand smoke. Secondhand smoke contains seven thousand chemicals, hundreds of them toxic—and at least seventy of them can cause cancer.[31] Approximately a million people die each year worldwide due to secondhand smoke.[32]

Then, there's thirdhand smoke. Thirdhand smoke is not actually smoke; it's the residue of cigarette smoke that creates the telltale smell on clothing or in a room. That residue alone can emit toxic chemicals.[33] Here's a tip: if you have the choice between a non-smoking room and a smoking room at a hotel, choose the non-smoking room.

7. Weight

In men and women over age fifty, excess body weight is associated with an increased risk of developing dementia. (In women, the association is more striking: older women who have significant belly fat have a 39 percent increased risk of developing dementia within fifteen years as compared to women with a normal waist size based on their height and build.[34]) Belly fat, or visceral fat, can release inflammation as well as disrupt metabolism and hormone release. Obesity can also cause blood flow and brain activity to diminish.[35] Low cerebral blood flow is one of the primary predictors of Alzheimer's disease; it is also associated with a range of other issues, from depression, ADHD, bipolar disorder, and schizophrenia to addiction. The prevalence of obesity may offer a clue as to why brain problems have been on the rise in the United States: 40 percent of Americans are considered obese, and another 30 percent are overweight—that's nearly *three quarters* of Americans.[36] Being overweight and/or obese is defined as having excess body fat, which is often measured in terms of one's body mass index, or BMI.

BMIs are calculated by taking a person's weight and dividing it by their height in meters squared. The "normal" range is defined as 18.5 to 24.9. A BMI of 25 to 29.9 is considered overweight, while 30 and above is deemed to be obese. The issue with BMI, however, is it does not distinguish weight that comes from fat from weight that comes from muscle. One could have a higher weight due to muscle mass, which is beneficial, and end up with a high BMI.

The waist-to-height ratio is more accurate than BMI in identifying obesity.[37] The waist-to-height ratio is calculated by measuring both the waist circumference and height in inches, dividing the waist circumference by height, and multiplying by 100. An individual is at higher risk of obesity-related diseases if they have a waist-to-height ratio of over 50 percent. A simple tip is to keep your waist to less than half your height. For example, someone 5 feet 4 inches (64 inches) should maintain a waistline smaller than 32 inches. When body weight is lowered to normal ranges as part of comprehensive intervention to improve health, there is also an improvement in memory and performance on cognitive tests.[38] (It's not just diet and exercise that lessen the risk of obesity: gut bacteria, hormone levels, environmental factors, and sleep can all play a role.)

That said, a higher body weight is not itself a conclusive indicator of poor health. What's most important is to not obsess over the number on the scale. Instead, think of being healthy as keeping your key numbers (such as cholesterol, blood pressure, blood sugar, and others we will discuss throughout this book) within the normal range.

A Few Last Words on Heart Health

Another way to improve your heart health? Attitude. People who are optimistic are twice as likely to have ideal cardiovascular heath as compared to those who are pessimistic.[39]

If we surround ourselves with optimistic people, we can lower our risk factors associated with Alzheimer's disease, dementia, and cognitive decline.[40] And having our own optimistic outlook can have an impact on our overall health, too! Optimists tend to live longer, with a 50 to 70 percent greater chance of reaching the age of eighty-five than those who are least optimistic.[41]

Studies suggest you can learn to be more optimistic—and that adopting a positive outlook about your ability to learn makes it easier for you to change.[42] In other words, you'll be more successful at becoming more optimistic when you're optimistic about learning to be optimistic. So, if you're pessimistic and say, "I can't learn to be more optimistic," you probably won't. You have to think you can get better at thinking that you can. That was exhausting, but you get the point.

If you are more of a pessimist, I have some good news for you, too—although I don't know if you will look at it as good news, being a pessimist. There are some benefits to being pessimistic at times. If we are too unrealistically optimistic, we can set ourselves up for disappointment.[43] Just like many of the factors we have discussed, it's about balance.

If you feel like you need a quick boost of optimism, here's a quick tip that has been shown to help people change their perspective. Think about somebody you have a relationship with that gives you mostly good feelings but also makes you upset, annoyed, or frustrated. You might say that's pretty much everyone I know! But pick one and then concentrate on the aspects of this person that you enjoy, love, and care about. Take a moment a couple of times throughout the day to think about that person and focus on the things you love about them, the things that you cherish and make you happy. Practicing seeing the good in others has been shown to make people more optimistic.[44]

I hope you leave this chapter with optimism. Since you know your brain and heart are connected, you know you can improve your heart health to protect your brain. Little changes can have a big impact. Now that we have covered the key connections between your immune system and heart, let's turn to the next piece of the puzzle: your gut.

CHAPTER 4

The Gut-Brain Connection

HAVE YOU EVER HEARD THE SAYING "LISTEN TO YOUR GUT"? WE ARE discovering a real connection between your gut and brain that shows just how wise that saying is. Turns out, your gut is a lot like a second brain. Your gut contains 500 million neurons, or brain cells, which communicate with your brain via the nerve that we mentioned in chapter three—the vagus nerve. This communication is called your gut-brain axis. This two-way connection is why when we are nervous, we get butterflies in our stomachs: emotions, stress, anxiety, and depression manifest in gastrointestinal symptoms. And what's happening in the gut can likewise impact mood. In fact, those with IBS or bowel issues are more likely to suffer from anxiety and depression.

It's not only neurons that make your gut like a second brain; your gut bacteria also play a vital role in this connection. So, let's talk about those bacteria. Before you close the book or skip ahead, I promise to make this the most exciting chapter you've ever read on bacteria (I know that's probably not a high bar).

I hate to break it to you, but you are half human, half bacteria. You are made up of about 37 trillion human cells, give or take a trillion. In addition, you have about 37 trillion bacteria living inside and on you.[1] All this bacteria weighs about five pounds. You might be thinking, *Wait, what? Half of me is bacteria yet it only weighs five pounds?* That's because a single human cell is much bigger than a bacterial cell. Thus, to be more accurate, *by number of cells*, you're about half human and half bacteria.

To delve deeper, I must do what I call "High School Biology in 30 Seconds." (Whenever I mention high school biology, I can practically hear the dread. But I will make this quick, and we'll cover a couple of key points.)

Bacteria are tiny, single-celled organisms. They come in all different shapes and sizes. Some of them are like little circles. Some look like squiggly lines. Bacteria sometimes have a tail. Inside your body they constitute the *microbiome*—a microscopic ecosystem of trillions of bacteria, fungi, and parasites of diverse species living together inside you. You can think of it like a forest filled with all different types of animals and plants living together. Your microbiome is critical for nutrition, immunity, and even human development. That's right, bacteria release chemicals that aid in orchestrating the developing fetus. Now that's some real symbiosis. An unhealthy microbiome is associated with diseases such as diabetes, autoimmune conditions, and brain dysfunction.

Bacteria and Viruses: What's the Difference?

Bacteria are not viruses. Bacteria are living cells that can reproduce by themselves. Viruses are not alive; they are just essentially a capsule with genetic material inside; they must get inside (infect) a living cell to reproduce.

What Are All Those Bacteria Doing?

All these bacteria must be doing something, right? We used to think they didn't do much good; they were either along for the ride or actively harming us. However, we now understand that the bacteria in your microbiome are critical for your mental and physical health in surprising and important ways. It's like that moment in the movie when you realize the person you thought was the villain is actually the hero.

Your skin is covered in more than 1.5 trillion bacteria right now. Before you get too grossed out, much of that bacteria is good for you. Bacteria that are necessary for us to survive and thrive we'll call "good" bacteria. These good bacteria on your skin eat mold. If it weren't for good bacteria on your skin, you would look like moldy bread. (OK, I realize it just got grosser. Sorry. Sometimes we just have to accept we're going to be covered in either bacteria or mold.)

These good bacteria live on top and in between your human cells, like friendly, peaceful neighbors. Even though we are covered in bacteria, much of the bacteria we are going to focus on in this book reside in the large intestine, or gut. The bacteria outside your gut obviously do play a role in health, but you have more bacteria in your gut than anywhere else on or in your body. And in terms of brain and immune health, the bacteria in your gut are the most impactful.

Your Agreement with Your Gut Bacteria

You and your gut bacteria have come to a mutually beneficial arrangement. You give them a place to live with just the right temperature, humidity, and lots to eat. In exchange, your gut bacteria help support your body functions, including metabolism, hormones, and immune function. For instance, does Tylenol work for you? The effectiveness of medications, in general, can depend on the types of bacteria living in your gut. Some people have bacteria in their guts that break the Tylenol into small enough pieces so the active ingredients can pass through the gut into the bloodstream and provide pain relief. Other individuals don't have those types of bacteria or don't have enough of them, so they don't receive the pain-relieving benefit.

Ten to thirty percent of patients with major depression don't respond to antidepressant medications. This has always been frustrating and puzzling.

A study from 2021 found that certain antidepressant medications such as duloxetine do not work in some individuals because of the bacteria in their microbiomes. The researchers discovered that certain types of gut bacteria essentially swallow the medication and don't allow the medicine to be absorbed into the bloodstream. In these cases, the therapeutic benefit of the medication can be significantly diminished.[2]

As another example, are you a chocoholic? It turns out if you're a chocolate lover, an element of this feeling is the result of certain types of bacteria in your gut releasing chemicals that impact your taste buds and induce cravings. If you're not a chocolate fan, you probably don't have these types of bacteria, or an abundance of them, in your gut. The next time you have a hankering for some chocolate, consider that it might be the bacteria in your gut having the craving.[3]

Those are just some examples of gut bacteria's surprising impacts. Gut bacteria also play a critical role in nutrition and metabolism. We eat a wide variety of food. Some foods like simple sugars are easily digested, and our bodies absorb the nutrients. Other foods with fiber, for example, are harder to digest, and the necessary nutrients harder to extract. In this case, the bacteria that live in your gut digest the food and free the nutrients so they can pass through your intestinal wall and get into the blood.

The bacteria in your gut do more than just digest your food, though. They also:

- Help manage your weight and metabolism (remember the connection between weight and brain health we covered in chapter three).
- Help balance your immune system (which we covered in chapter two).
- Make certain vitamins, essential micronutrients our bodies cannot make themselves (notably B-group vitamins, covered in chapter three).
- Play a surprising role in our brain health (which we're about to explore!).

Where Do We Get Our Bacteria?

How we are born plays a role in establishing our initial colonies of bacteria. Babies born vaginally obtain much of their initial bacteria in the birth canal.[4] Babies born by C-section receive their initial bacteria from the bacteria present in the delivery room. Over the next few years, the infant's body hosts a Wild West–style fight to determine which of the different strains and species of bacteria will inhabit the gut. The health impacts of vaginal versus C-section births are not completely clear, but a 2020 study found that C-section babies have a higher risk of developing rheumatoid arthritis, celiac disease, and inflammatory bowel disease even up to forty years after birth.[5] However, another 2020 study found that breastfeeding restored missing gut bacteria in babies born by C-section and lowered infection risk in early life.[6] Breast milk is filled with good bacteria.

Factors that determine which bacteria stay in the gut include living in rural or urban environments, outside playtime, medications, pets, and diet. Throughout life, the types of bacteria can be altered by changes in environment and diet. These alterations can either be beneficial or detrimental to health.

Bacteria and the Gut-Brain Axis

Neurotransmitters such as serotonin and GABA are produced in the gut by bacteria—actually, your gut produces the vast majority of your body's serotonin. In the brain these neurotransmitters regulate emotions such as happiness, fear, and anxiety. Thus, the types and quantities of gut bacteria can impact emotional states. Gut bacteria also produce chemicals in the gut such as butyrate, which impacts brain and immune function. For example, butyrate has anti-inflammatory properties and protects brain cells.[7]

The first indication that gut bacteria impacted the brain came from a groundbreaking study with mice published in 2011, suggesting that certain types of gut bacteria might have a more calming effect on the brain, while other bacteria can play a role in anxiety.[8] Mice have different personalities. Some mice are adventurous and daring, and other mice are shy and timid. Scientists wondered, *What if we took gut bacteria from an adventurous mouse and transplanted them into the gut of a shy mouse, and transferred gut bacteria from the timid mouse to the adventurous mouse?*

When they did this, the daring mouse displayed shyer behavior, and the shy mouse became more brazen than expected. The researchers also discovered that when the vagus nerve was severed *and* the gut bacteria were swapped, the personalities of the mice did not change. This indicated that the vagus nerve was the essential method of communication between gut bacteria and the brain.

Think of your vagus nerve like a guitar string. If that string is plucked quickly, it can cause stimulating feelings that can lead to anxiousness. On the other hand, that string could be calmed down with a slower rhythm. Different species of bacteria release stimulating or calming factors impacting the vagus nerve. This is just one way in which what is happening with your gut bacteria can impact your mood. (Subsequent studies have uncovered a link in humans between the types of bacteria in the gut and brain disorders such as anxiety, depression, and schizophrenia.[9])

How do the gut and brain communicate? Normally, something called the blood-brain barrier protects the brain from factors circulating in the blood. Think of it like a security fence around your brain so that not everything that enters into your bloodstream impacts brain function. Factors released by the microbiome *can* penetrate the blood-brain barrier, thus impacting brain function and metabolism and how discriminating the barrier is at letting chemicals, nutrients, and a variety of factors into the brain.[10] The microbiome's ability to pass through this barrier highlights the power of the connection between our guts and brains.

The gut-brain connection is also a two-way street—information travels in both directions. Anxiety and stress, or anxious and stressful thoughts, cause the release of cortisol from the adrenal gland into the bloodstream. Levels of the hormone can build up in the gut and promote the growth of harmful bacteria, which send signals via the vagus nerve back to the brain, creating a vicious cycle. The excess cortisol in the gut excites the vagus nerve, stimulating the release of still more cortisol, promoting feelings of stress and anxiety. In some cases, treatment for gut issues can involve stress management, and treatments for stress and anxiety can involve interventions in gut health. In upcoming chapters we discuss techniques to optimize your gut health to improve brain health and vice versa.

What's most important for this chapter, though, is that through nerve connections, neurotransmitters, and key chemicals, what is happening in the gut can impact mood, memory, and how we age. Gut bacteria not only aid digestion but also enhance immune function, prevent obesity, and even play a role in conditions such as Parkinson's, Alzheimer's, autoimmune disorders, anxiety, and depression. For example, a study from 2020 confirmed that some people with Alzheimer's have different types and combinations of bacteria growing in their gut compared to those without Alzheimer's.[11] There is evidence that certain types of harmful bacteria release toxins and plaque-like material that can leave the gut and make their way to the brain, where they cause the kind of damage associated with dementia and Alzheimer's.[12]

When Bacteria Are Out of Balance: Inflammation and Immune System Issues

In a healthy individual, there is a balance between good and bad bacteria. Some bacteria are dangerous, but they're a tiny proportion of all the bacteria living inside you. And as we continue to learn about the functions of different bacteria, we realize that some initially thought to be bad actually benefit us in the right amounts as they coexist with other species of bacteria. If the overall balance of good and harmful bacteria becomes disturbed, a large variety of metabolic and autoimmune diseases can ensue. The immune system helps keep dangerous bacteria from overgrowing and causing damage.

Just like in real estate, sometimes it's all about location. What we would call good bacteria in one part of your body can be harmful when they're in the wrong place. (An example of this is bacterial overgrowth in the small intestine. Certain types of bacteria in the large intestine can be helpful, but if these bacteria start growing upstream in the small intestine, this can lead to stomach aches and malabsorption of nutrients. This condition is called small intestinal bacterial overgrowth syndrome, or SIBO. A variety of underlying conditions, such as irritable bowel syndrome, Crohn's disease, intestinal surgeries, cirrhosis of the liver, alcoholism, and fibromyalgia, raise the risk of SIBO. These conditions can increase the risk of bacteria either migrating from the large intestine to the small intestine or overpopulating the small intestine. To highlight the complexity of the different types of bacteria, a study identified 141 different strains of bacteria that can cause symptoms of SIBO.[13] There is evidence that SIBO can increase the risk of "brain fog," which is a general term for difficulty concentrating, confusion, and memory problems.[14])

Inflammation can cause a variety of symptoms, from pain and soreness to brain fog. And what can be the culprit? Yet again, bacteria play a role. (If you answer pretty much any question posed in this chapter with "bacteria," you have a good chance of getting it right.) Bacteria line your gut and act like a bouncer in an exclusive club, only allowing nutrients to pass into your bloodstream; these are "good" bacteria. Some types of gut bacteria aren't as discerning.

An unhealthy gut can have an overgrowth of "harmful" bacteria, which excrete chemicals that increase inflammation and damage the lining of the gut, until it is no longer an effective barrier and permits toxins and waste to leak into the body. A leaky gut can also allow large fragments of undigested food to enter the bloodstream. The immune system recognizes these toxins, wastes, and food fragments as foreign and tries to eliminate them, causing inflammation. This inflammation is a factor in autoimmune diseases such as lupus, multiple sclerosis, rheumatoid arthritis, heart disease, diabetes, and obesity. Inflammation can spread and also attack the brain, leading to premature aging of the brain and increased risk of brain dysfunction. Just imagine the brain being attacked and you'll get a sense of how inflammation can cause serious damage to the brain. The microbiome is emerging as a critical factor in Parkinson's disease. Gastrointestinal symptoms often precede the symptoms of Parkinson's. These symptoms can be related to an imbalance in gut bacteria that can contribute to inflammation in the gut and the brain in Parkinson's disease.[15] Evidence also suggests that specifically in Alzheimer's disease, the microbiome can stimulate the microglia, thus increasing inflammation in the brain, which can play a role in cognitive decline.[16]

We used to think about the immune system as a complex army designed to protect you against what's not you. How do we fit this into our new understanding that about half of you is not you because half of you is bacterial cells? How does your immune system know not to get rid of the bouncer-like bacteria because they're actually keeping toxins from entering your bloodstream? How does your immune system know which bacteria or viruses to eliminate and which to protect?

If you are a binge watcher of television shows such as *Law and Order* or *The Blacklist*, you know the detectives often use an informant to help them solve crimes. Think of the detective as the immune system and imagine the informant as the bacteria. Bacteria teach and train the immune system. Good bacteria will tell the immune system which other good bacteria to preserve and which harmful bacteria to remove. On the other hand, bad bacteria can give incorrect information and tell the immune system to remove good bacteria and protect bad bacteria. Some bacteria you just can't trust!

Bottom line: we want the good bacteria that protect us to flourish in our guts, and we want to minimize the growth of harmful bacteria. A healthy gut is filled with diverse species of bacteria. By diverse, we mean lots of different strains of bacteria. As we age, we tend to lose our gut microbiome's diversity, which can wreak havoc on our gut, brain, and immune health. We create a diverse microbiome through what we eat. I go into detail on how to do that in chapter fourteen, but for now, the important thing to remember is we need to embrace the health of our bacteria as a critical piece of our mental and physical health.

Taking Care of Your Gut Bacteria

Though your diet makes a big difference in the health of your microbiome, nurturing gut bacteria isn't just about food. For example, antibiotics are amazing modern miracles of medicine, but they can kill bacteria indiscriminately, knocking out both the good and the bad, so it is critically important to use them only when necessary and only as directed by a physician. This is the same for all medicines; we want to use them when we need them but not when we do not. If you're prescribed antibiotics, it is essential to ask the prescribing physician if probiotics should be taken with the antibiotic to replenish lost good bacteria. It is also important to consider including probiotic and prebiotic foods in the diet while on antibiotics; a study found that taking probiotics with antibiotics reduced the risk of stomach issues related to antibiotic use.[17] (If you're not yet familiar with pre- and probiotics, worry not; they're covered in chapter fourteen.)

Likewise, take care around some kinds of soap. Antibacterial soaps can also kill off good bacteria. We need to wash our hands, but antibacterial soaps are unnecessary (unless we're about to perform surgery). Just stick to old-fashioned soap and water.

I hope you leave this chapter as fond of bacteria as I am. Or at least fonder than you were before. The key take-home message is that bacteria are a piece of the puzzle to our mental and physical health. While it's not the whole puzzle, it's a very important piece. So be good to your gut bacteria. They are doing amazing things to protect and nourish you. Feed the good bacteria. Starve the bad bacteria. Remember, you and your bacteria are in this together. I hope you and your gut bacteria enjoy your next meal!

CHAPTER 5

How Memory Works

STEPHEN WILTSHIRE CAN TAKE A FIFTEEN-MINUTE HELICOPTER RIDE over a city he has never been to with no camera or notes and, after that short ride, paint a detailed skyline of the city with incredible accuracy.[1] When Stephen was briefly flown over the Pantheon, he was able to include the exact number of columns in his rendering. Stephen is a savant, which means he has superhuman memory capabilities but at the same time has difficulty with day-to-day memory or communication. How Stephen does *his* memory-based drawings is still a mystery, which highlights the complexity of the human brain and how much we are still learning about its capabilities. But memory prodigies have helped us make major strides in our understanding of how memory works and how to improve it. There are also memory champions such as Alex Mullen, who took twenty seconds to memorize the order of a deck of playing cards. When Alex was done with that, he memorized the order of three thousand numbers.[2] You may be saying, "I don't really need to memorize decks of playing cards anytime soon (or at least until my next trip to Vegas), so what do Stephen and Alex have to do with me? I just want to remember where I put my phone." But understanding how the brain works and how memories are formed allows us to swipe some of the clever tricks of memory champions to remember day-to-day things, from where you parked your car or where you put your keys, to people's names and where that important piece of paper is you are turning your house upside down to find. These techniques can also be used to improve memory at school or work. Memory tricks are based on the science of how memories are made.

Your Remarkable Memory

Small, often overlooked things you do every day highlight how incredible your memory is. Here's a quick example. If you know how to touch-type, you can likely do this very easily and quickly with a flurry of finger movements. But can you tell me which fingers push specific letters? Most people can't. This is an example of how your memory is great at knowing things you can't articulate.

Keeping your brain young lets you continue to make new memories and access old memories. You might think that your memory is like a video camera recording; when you want to revisit a memory, all you have to do is hit replay. Although this is a common intuitive assessment of memory, it isn't how memories are made and retrieved.[3] And though we often tell ourselves that we'll "just remember" something important, this won't necessarily make a memory that sticks; it often takes more effort and work. By understanding the key brain structures and how they work, we can truly improve and maintain our memory.

Three Key Aspects of Memory

There are three types of memory: sensory register, short-term, and long-term memory. Sensory register is when the brain collects information from our surroundings through the senses. An example of this fleeting memory is when you watch the fireworks and you see the streaks of light a moment after the firework disappears. This type of memory is fascinating, but for the purpose of boosting memory, we'll focus on the other two.

We make and remember memories through processes called encoding, storage, and retrieval. I have found in speaking on this subject that there are key aspects of memory that can be remarkably helpful and meaningfully improve our recall. They are:

1. True focus or attention
2. Short-term memory
3. Long-term memory

If we skip or ineffectively pass through any of these, we will likely not remember the information.

1. True Focus or Attention

As you read or listen to this page, there is a lot of stimulation happening around you. Whether you are alone in a peaceful cabin miles from civilization or reading this on a crowded bus trying not to smell the person next to you, a lot is going on that your brain (and nose) must deal with. If you are sitting or standing, your brain must process the pressure of your feet against the floor or the feel of your chair. Your brain is also registering the temperature of your environment as well as even the slightest background noise. Right now, your brain is suppressing much of your conscious awareness to just focus on this information. If it weren't—or if it were focused on sensory information that should be background noise—you'd miss the opportunity to make a memory with meaningful information. This is like when you are trying to study for a test but your brain decides this is the perfect time to analyze what your carpet looks like when you really focus on it. It all comes down to the fact that we cannot make a conscious memory if we don't block out all the unimportant stimuli around us and pay attention through true focus. In the process of making a memory, true focus, or attention, is a step between sensory register and short-term memory. True focus is what signals to your brain that it should even try to make a memory. And it's critical to understand that the brain's capacity for focus is limited.

Take a look at almost any photo of Steve Jobs from the 1990s to 2010. He wore pretty much the same thing every day: blue jeans and a black turtleneck. Why would a man who could afford a closet full of choices pick the same outfit every day?

Steve Jobs had some insight into his prefrontal cortex (PFC), the part of the brain that controls focus. Back in chapter one, we took a brief look at the parts of the brain; you may remember reading about the cerebral cortex (the gray matter on the surface of the cerebrum) and the frontal lobe. The PFC is part of the cerebral cortex; it covers the front part of the frontal lobe. Your PFC sits right behind your forehead.

The PFC—and thus your ability to focus—is a limited resource. Think of it like your cell phone battery after you unplug it from your charger: you start the day at full power but the battery drains as you use it. Jobs realized this and decided that he would not waste any of his mental battery on deciding what to wear. By wearing the same thing, he saved his energy and his PFC for things he felt were more important. I am not suggesting we all wear the same outfit each day (though, like Jobs, there are people who do and swear by the mental energy it saves);[4] however, it *is* important to determine which distractions and decisions you can eliminate so you don't waste the precious resource of focus. Create a morning routine, make your to-do list the night before, or pick out your clothes before you go to bed to save your energy for focusing.

2. Short-Term Memory

Have you ever met someone and said to yourself, "I'm going to remember this person's name"? Then they tell you their name and it vanishes into thin air. When people start to experience these normal memory lapses, they often begin to panic. Sort-term memory snafus are based in the part of your brain called the hippocampus. (Humans actually have two hippocampi, one in

The Dopamine Squirt

Want to boost your focus? Give your brain a squirt of dopamine. Dopamine is a neurotransmitter—a chemical released by your brain cells that transmits messages to other brain, muscle, and gland cells.

The brain releases dopamine when we are surprised or confronted with something new and outside our previous experience. Because dopamine feels good, our brains focus better on the task, chasing more dopamine. Presumably, our ancestors who did not pay attention to new or unexpected things—like the tracks of a predator near the village—could be injured or miss out on an opportunity to learn something important.[5] The dopamine squirt put them on alert. Now most of us are unlikely to run across a tiger on our morning stroll, and if you do, you might want to consider changing your ZIP code, but our brains still squirt dopamine when we encounter something new. Newness can be as simple as going for a walk on a new street, cooking a new recipe, taking up a new hobby, or singing a new song. Try a change of scenery: work in a different room if you work at home or find a new coffee shop. This is why we often have great ideas on vacation, and why a trip can freshen up a relationship—the dopamine release causes greater focus on a problem, or on each other.

each of the temporal lobes.) The human brain has evolved over hundreds of thousands of years to keep itself tidy and organized and to not accumulate too much useless, distracting information. New information goes first to the hippocampus; it's like a waiting room where your brain decides what's important or not. Important information is transferred to other areas of the brain for long-term memory storage. If your brain determines the information is not important, it will be discarded, and you will not remember it. If your brain didn't discard information at this step, it would simply fill up with too much useless stuff and wouldn't work efficiently.

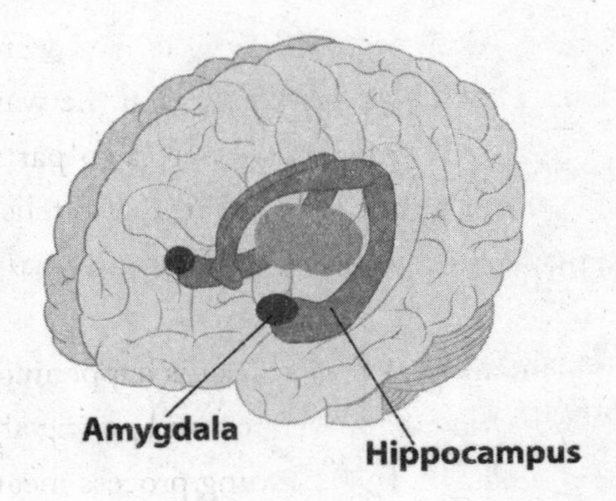

Short-term memory doesn't retain something that happened yesterday or a week ago. *Your short-term memory is something that only lasts seven to twenty seconds.* That's right: you can hold new information in your brain for about seven to twenty seconds. We know this because of one of the most famous patients in brain science, Henry Molaison, better known as H.M. Born in 1926, H.M. grew up as a normal child until the onset of epilepsy (attributed to a bicycle accident at age seven). He started having seizures that were initially minor but became more and more severe, until they occurred so often and became so intense that he could not leave his home. Doctors tried every therapy, every medicine available at the time. Nothing worked. H.M.'s only hope was a surgery that was performed on him and never performed again: In 1953, when he was twenty-seven years old, doctors removed his hippocampus, which was believed to be the source of his seizures. This dramatically reduced the number of seizures but left H.M. with a devastating side effect: he developed what is called anterograde amnesia. He could still access memories he had from before the procedure, yet he could not form any new memories after the surgery. In fact, if you met H.M., you would have to reintroduce yourself to him every seven seconds. That was how the scientific community first understood that the hippocampus is the part of the brain responsible for learning, memorizing, and taking in and processing new information.

You may recall a time when we had to remember phone numbers. If you didn't have a pen and paper to write down the number, you would repeat it to yourself. Every time you repeated the digits, you were giving your hippocampus more time to hold that information in the waiting room of your brain. By repeating the information, you were also partaking in a process called rehearsal. In memory formation, rehearsal also tells the brain that this information is important enough to move from a short-term memory to a long-term memory.

Let's take a moment and talk about what is happening inside your brain. The sensory information you are focusing on and paying attention to needs to be encoded to form a memory. The encoding process means physically turning this information into brain cell connections. When you make a memory or learn something, a subset of your eighty billion brain cells makes a new connection and encodes the information. This is like laying down train tracks that store the information. You will eventually revisit these same tracks or brain cell connections to retrieve or remember the memory.

Working Memory

Working memory is a specific type of short-term memory. It is like a notepad in your brain, where you are working with information you just learned. For example, while you're watching a complicated detective movie, you're keeping all the plot twists in mind.

I bet you have experienced firsthand what happens when the encoding process is interrupted. As an example, have you ever found yourself in your bedroom, and you know there was something you wanted there, but you can't remember what the heck it was? The next time this happens, see if you can remember how much time elapsed between when you had the idea of what you wanted and when you arrived in the room. I bet it probably took you about seven to twenty seconds—in which time you probably got distracted by something (maybe you got a text or started thinking about what to eat for lunch). Because if you got distracted and thought about something else in that time, what would your hippocampus do with the idea of that thing you wanted in your bedroom? That's right—it would delete it.

(As an aside, you may *also* have the experience of "remembering" what you forgot when you go back to what you were doing just before. Say you had been in your kitchen, thought of something you needed in your bedroom, walked to your room, couldn't remember why you were there, and went back to the kitchen—whereupon you remembered, *Oh, yeah! I wanted to change my socks*. Scientists aren't totally sure why, but they believe that the phenomenon is a function of the brain having that same idea or thought again by retracing steps. There was even a unique study that found when people literally walked backwards, they improved their short-term memory.[6] It's possible these people were literally retracing their steps. Weird, right?)

So, the key to optimizing your short-term memory is to slow down, eliminate distractions, and do one thing at a time when trying to memorize or learn. Practice taking at least seven uninterrupted seconds to focus on what you are doing. In this time, you are rehearsing the information and thus taking part in the encoding and storage process of memory formation. We live in a world where we often don't take even two seconds to absorb our environment, yet we wonder why we keep forgetting. People are surprised how often they remember things better if they utilize the seven-second rule and just focus for a few extra seconds. For example, if you are reading and come across a piece of information you want to remember, consider re-reading that sentence a few times without distraction.

Filtering Matters

Hyperthemesia is an extremely rare condition in which individuals have what is called superior autobiographical memory.[7] Reports suggest that only about sixty people have ever been identified with this condition (including, just as an interesting bit of trivia, the actress Marilu Henner).[8] Ask one of these people what happened on a specific date, such as September 1, 2017, and they will tell precise details about their life, such as what they had for each meal that day, who they spoke to, and specifics on the conversations.

It might sound great to remember everything so you could ace every test and be a superstar at work, but many of the people with this condition say it makes it difficult to focus. Filtering is an important part of memory. We want to filter out the unimportant, mundane aspects of life so we can focus on the key things we want to remember. There is still a lot to learn about this condition, but brain scans have revealed that individuals with hyperthemesia tend to have an enlarged and overactive hippocampus.[9] We don't want a shrinking hippocampus, but these findings suggest that balance is key. These cases are fascinating, and insights from these individuals could uncover ways to treat memory loss. There is evidence that the ability to forget is important

The Seven-Second Rule

Someone once told me, "Forget my name and I will never forgive you. Remember my name and I will never forget you." Now if I could only remember the name of the person who told me that.

Just kidding.

I remember his name. It's Paul, and I remember his name because of this brain science-based trick I am about to show you.

Let's discuss that moment when you meet someone and you think, *I am going to remember their name. This time I am going to do it!* They tell you their name and it's gone.

When you find yourself thinking, *What was that person's name?* take a minute to reflect. Did you really take seven seconds to stay focused on what you were hearing? Or was there a distraction? Or were you multitasking, texting, cooking, or wondering, *Are they looking at my mouth? Do I have food stuck in my teeth?*, all while trying to have a conversation?

The next time you meet someone and you want to remember their name, imagine yourself writing their name on their forehead. Why does this silly trick work? In the seven to ten seconds it takes to "write" their name, you have convinced your brain this person's name is worth it.

for remembering important information.[10] For now, let's embrace some forgetting! I for one am thankful that I have forgotten much of high school.

3. Long-Term Memory

Once the information leaves the hippocampus, where does it go? It's time to further encode and store your memories.

As discussed in chapter one, you can think of your brain like a bank account: when you want to make new connections, it's like building up a savings or reserve for when we (inevitably and normally) lose some connections down the road. But out of 100 trillion connections in your brain, how do you find the connection that houses the memory you are trying to remember? The process of remembering is called the retrieval stage in memory formation. Let's think about how you retrieve a memory.

Take a moment and think about someone specific in your life. Pick a friend, a relative, an acquaintance, a coworker, even an ex, if you want to go down that road. Is there a specific section of brain cells that we could point to that holds the memory of that person? Could we completely erase the memory of that person if we somehow removed those brain cells? While movies like *Eternal Sunshine of the Spotless Mind* might suggest that we can do just that, to be blunt, we can't. (Unfortunately, for this reason I won't be able to simply erase the memory of my prom night.)

As mentioned earlier, your memories are stored in the connections between brain cells. And the memory of that person is stored throughout your brain, not just in one location. The way the person looks is stored in your visual cortex, the way the person sounds is stored in the auditory part of your brain, the way the person smells is stored in the olfactory part of your brain. How you feel about that person is stored in the emotional part of your brain, called the amygdala. Each sensory and emotional experience lives in a different and distinct region of the brain, and all your memories are broken up into separate components and stored throughout those regions. Amazingly, when you think of your friend (or relative or coworker or ex), you pull all this information together into one cohesive memory.

You can use the knowledge that information is stored throughout your brain to your advantage. In order to remember information, strengthen the connection between the brain cells that house the memory by reviewing or practicing the information. For example, if you learn a tune on the piano, you make a connection. Every time you practice the song you learned, you make that connection stronger. When you go for months without practicing, those brain cell connections get weaker. (This is why it is so hard to break some habits—the brain cell connections are still there even if they are not as strong. It's also why it's hard to form a habit, because you have to make these new connections and that takes work. We'll explore some other ways to strengthen the connections between cells and make habits in later chapters.)

Have you roamed the parking lot looking for your car? Next time you park your car and want to remember where it is, stop for a moment and say, out loud, something that addresses the location. For example, if it is in a multilevel garage, say, "I parked in level 2, section B." (Don't worry that people will think you're talking to yourself; everyone will assume you are on your Bluetooth.)

Why does speaking out loud make a difference? Your memories are like nuts that squirrels hide for the winter. The more places they are stored, the more places there will be to access later, increasing the probability of successfully finding a nut—or recalling a tidbit. The part of your brain that is involved with speech is different than the part of your brain that's involved in hearing—but *both* those parts of the brain have memory storage capabilities. If you say something out loud, you've just stored that memory in the part of your brain involved with speech *and* the part of your brain that's involved in hearing. You can use this trick to remember where you put your keys, someone's name, and many other things you want to remember. Just say it out loud.

Practice Almost Makes Perfect

Remember, to really make the memory stick, you need to review the information. When we review, we run electrical stimulation over the connections that house the memory, making that connection stronger. If we don't review it, the brain figures this information is no longer needed and it is harder to remember the information. One way to think of this is that when a connection weakens, it rusts like old train tracks until it crumbles and makes it difficult, or in some cases impossible, to travel that road.

But do we ever really forget something we have learned? This is not completely understood, but there is evidence that it is easier to relearn something you forgot than to learn something completely new. Part of what is happening is those brain cells that house the information can still be present; they are just too weak to remember. The relearning process strengthens those brain cell connections, which can be easier than forming a brand-new connection.[11]

Let's review the memories we just made together to make the connections stronger:

True focus or attention: There is a lot of useless sensory information that we need to filter out in order to remember what is important. The prefrontal cortex, the part of your brain behind your forehead, needs to be engaged to focus on important information. A dose of newness releases the chemical dopamine in the brain, which can boost focus.

Short-term memory: After we focus, the brain continues to determine if information is worthy of remembering by briefly storing it in the waiting room of the brain, the hippocampus. Information waits there for about seven to twenty seconds while the rest of your brain determines if the information is worthwhile. We can remember more of the things we want to recall by simply focusing on the information for an uninterrupted seven to twenty seconds. This helps convince the brain the information is important and should not be thrown away. During this process you are storing and encoding information by creating new connections between your brain cells.

Long-term memory: Once information is worthy, it leaves the hippocampus and is encoded and stored all throughout the brain. We store the memories in newly formed connections between brain cells. By storing information in multiple parts of the brain, we have a greater chance of recalling it. For example, taking an extra few seconds to say the information out loud can save minutes struggling to recall information later.

Now that you've learned about how your brain developed, how it interacts with other body systems, and how memory works, it's time to shift gears. In the next part of this book, we'll talk more about common brain health issues and the surprising factors that age the brain. Since we've just covered making and using memories, we'll first look at one of the most worrisome: memory loss and dementia. (But before you turn the page, remember there's a light at the end of the tunnel—these common issues can be prevented, and we'll get to how in part three.)

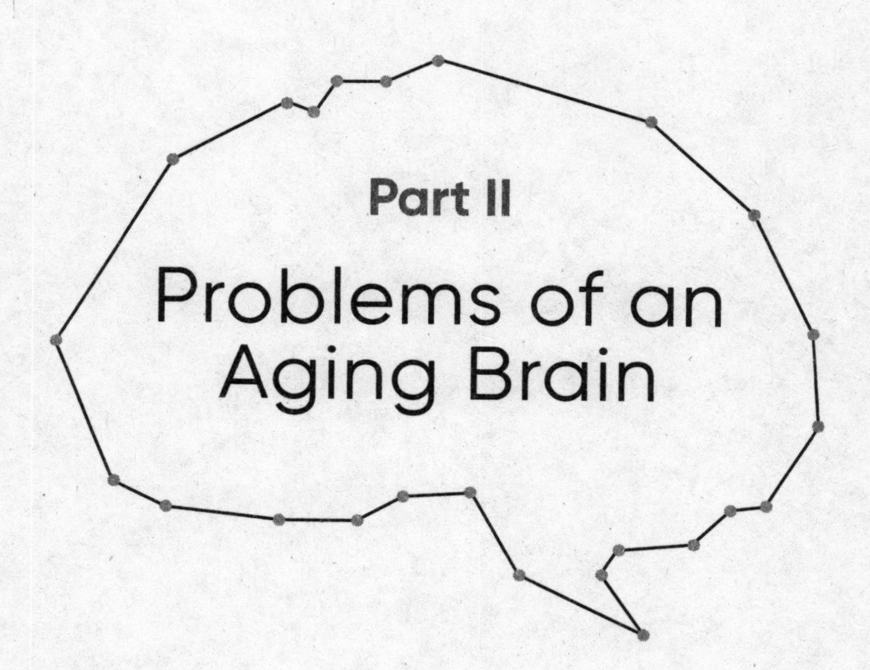

Part II

Problems of an Aging Brain

CHAPTER 6

Memory Loss and Dementia

THE WAY WE THINK ABOUT ALZHEIMER'S AND DEMENTIA HAS changed dramatically. We used to think dementia was something that came on suddenly and was out of our control. Now we know more.

Imagine yourself at the beach. Those waves rolling in are the dementia symptoms. If you look past the waves, it looks calm—but the waves are there, just beneath the surface. Many of those below-the-surface waves have been traveling for hundreds of miles, gathering power. Loss of mental function is like a wave crashing onto the shore: we aren't aware of it until it knocks us off our feet. Like those waves beneath the surface, changes begin in the brain and body *decades* before we ever see a symptom.

As I mentioned in this book's introduction, rates of dementia are rapidly increasing. And many of us have witnessed firsthand the devastation of memory loss, which takes our loved ones from us too soon. But there's plenty of reason to be hopeful. As we'll see later in this chapter and in part three of this book, we can take steps to protect our memories. So that's why, before we talk about more prevention, it's worthwhile to spend some time getting to know more about memory loss and dementia.

Memory Loss Without Dementia

Before we discuss dementia (and especially Alzheimer's disease), there is a form of age-related memory loss that is not as severe as dementia. This is called mild cognitive impairment (MCI). Between 12 and 18 percent of those over the age of sixty suffer from MCI—which is *not* a normal part of the aging process.[1] While it can sometimes be tough to tell the difference between "normal" brain aging, MCI, and dementia, there are key distinctions. MCI is a type of memory loss and impaired mental function that is noticeable; someone with MCI might have difficulty thinking during a conversation, get lost in a familiar area, or have trouble completing a routine task such as making a travel reservation or paying a bill. They might also have problems with balance and coordination. However, these issues are less severe than what we often see with dementia; someone with Alzheimer's (or another form of dementia) will struggle more with regular tasks and might also show signs of impaired judgment.

There is not one lone cause of MCI, but it can be caused by brain aging. Those with MCI often have higher levels of waste, plaques and tangles, and/or reduced blood flow to the brain. In some cases, this reduced blood flow can be caused by small strokes. There can also be damage to the hippocampus whenever there is reduced blood flow to this region of the brain, no matter what the cause.

MCI might not be as severe as dementia, but it can lead to it. Although not all those with MCI will develop dementia, 10 to 15 percent do. There is a need to raise awareness of MCI and not consider these memory struggles an inevitable part of aging. The root cause needs to be addressed, diagnosed, and treated if possible. The lifestyle factors discussed in section three of this book can protect the brain from aging and MCI.

Understanding Dementia

While the two conditions are frequently confused, Alzheimer's and dementia are *not* the same.

Dementia is a set of symptoms that includes memory loss, having trouble making decisions, and personality changes that interfere with getting through the day.[2] Alzheimer's is a specific disease that *causes* dementia. It is the most common cause of dementia; estimates suggest that Alzheimer's disease is the root cause of between 60 to 80 percent of dementia cases. However, there are many other known types of dementia. The box that follows breaks those down.

Vascular dementia

Vascular dementia is a brain dysfunction caused by a lack of blood and oxygen to the brain. One of the most common causes is stroke. Even small, or what are called "silent" (undetected), strokes can lead to dementia. The symptoms of vascular dementia are often related to where in the brain the stroke occurs. Some of the most common signs include difficulty making decisions and poor judgment.[3] As discussed in chapter three, the connection between the heart and the brain is critical. One must have a healthy heart to have a sharp brain.

Mixed Dementia

Mixed dementia is a combination of two types of dementia. The most common combination is Alzheimer's disease and vascular dementia. Multiple causes of dementia can make the symptoms worse. A study done at the Rush Institute for Healthy Aging found that half of the brain autopsies of Alzheimer's patients also showed evidence of vascular dementia.[4] In most cases, this secondary cause of dementia was not diagnosed during the subject's lifetime. It is important to determine if there are multiple root causes or mixed dementia present, as it can inform treatment.

Lewy Body Dementia

A feature of Lewy body dementia is the formation of waste deposits in the brain that build up in the cortex, wreaking havoc on cognitive function, attention, and decision-making. These deposits are different proteins than those found in Alzheimer's, but like Alzheimer's, they clump together and can cause dementia. Lewy body dementia can cause hallucinations, sleepiness, and difficulty with movement. There is a dire need for more research on Lewy body dementia and its causes and interventions.

Parkinson's Disease Dementia

Parkinson's is another disease that causes the formation of trash in the brain. Initially, the waste disrupts the brain's ability to make dopamine, a crucial chemical involved in controlling movement and paying attention. This lack of dopamine is responsible for tremors present in Parkinson's disease. Depending on certain factors, including duration and age of onset, approximately 50 to 80 percent of Parkinson's cases progress to dementia.[5] An emerging area of research is the connection between the gut and the brain in Parkinson's. In some cases, individuals with Parkinson's experience digestive symptoms years before developing any cognitive difficulties or tremors. There is evidence that in some cases, Parkinson's might begin in the gut.[6]

One way to think about the difference between Alzheimer's and dementia in general: say you have a runny nose. The runny nose is a symptom, but the cause could be a cold, allergies, or a reaction to the temperature of the room. Likewise, *dementia* describes a group of symptoms, but the cause could be Alzheimer's, vascular dementia, another form of dementia, or something else entirely. Because Alzheimer's disease is the most common cause of dementia, this chapter will discuss it in more detail.

Sorting Dementia Fact from Fiction

Myth: There is no way to prevent or reverse dementia.

Fact: In rare cases of genetic Alzheimer's and certain physical conditions like Lewy body dementia, this is true. However, approximately 10 to 20 percent of all cases of dementia—those caused by a hormonal imbalance, infection, vitamin deficiency, a side effect of a medication, or other factors—are often partially or completely reversible.[7] The range of success in partially or completely reversing symptoms of dementia is based on multiple factors, including age of onset and the root cause of the dementia.

One study used the mnemonic DEMENTIA to classify causes of specific types of dementia that are either partially or completely reversible:[8]

Drugs
Emotional
Metabolic
Eyes and ears declining
Normal pressure hydrocephalus (fluid buildup on the brain)
Tumor or any lesion in the brain
Infection
Anemia

A key take-home message is that we always want to figure out the root of dementia because in a significant number of cases, that root can be treated.

A Quick Guide to Alzheimer's

Alzheimer's is a progressive disease that worsens over time. In the past, Alzheimer's was only definitively diagnosed after death, at autopsy. Now a combination of an evaluation by a physician, brain scans, and specific tests can provide a diagnosis. Since Alzheimer's takes years to develop, early diagnosis is important.

There are multiple factors involved in the development of this disease, but a key factor is thought to be the formation of plaques and tangles—the brain trash I introduced in chapter one. Tau protein tangles and beta-amyloid plaques prevent communication between brain cells during Alzheimer's disease. That said, there is disagreement among researchers about whether the plaques and tangles are *responsible* for memory loss or if they are *by-products* of damage to the brain.[9] For instance, scientists studied a community of nuns in Minnesota.[10] Sister Mary was 101 years old. She was mentally sharp. Crosswords and puzzles were no match for her. So, when she died at 102, scientists were surprised to find upon autopsy that her brain was filled with plaques and tangles. Sister Mary was not the only case. Other studies have found individuals whose brains were littered with significant amounts of trash but who had no memory loss or loss of mental function. So, while brain trash plays a role in dementia, in some cases they don't seem to be enough, or the only factor or tipping point.

Other factors involved in Alzheimer's disease include inflammation (as we'll discuss in chapter eight), metabolic dysfunction (covered in chapters three and seven), and vascular issues (as covered in chapter three). Through a complex mix of these processes, brain cells age, become damaged, and cause brain shrinkage. This loss of brain volume devastates memory; in particular, the hippocampus is often damaged in Alzheimer's and dementia. If you recall this key brain structure's role in memory, explored in the previous chapter, you'll probably see why individuals with dementia will often be able to recall events that happened decades ago but cannot remember new information—due to damage to the hippocampus.

Another factor can be genes. Only 1 to 5 percent of all cases of Alzheimer's are strictly genetic. That means that most of us are *not* destined to develop Alzheimer's based on our genes. Those 1 to 5 percent of strictly genetic cases are due to *deterministic* genes: rare genes that, if inherited, mean that someone will definitely develop early onset Alzheimer's (that's when the disease sets in between one's early forties and midfifties). There is a documented exception to this: A 2019 study examined a woman who inherited the deterministic Alzheimer's genes that lead someone to develop dementia in their forties or fifties but mysteriously made it to her seventies before developing mild dementia.[11] Researchers analyzed her DNA and discovered she had another rare gene mutation that seemed to protect her. While we need to proceed with caution, because this is just one study of one woman, it does provide hope for those with deterministic genes. Perhaps by further understanding these gene relationships, there could potentially be treatments for those with these determinist genes.

There are other genes, such as versions of APOE, that can raise one's *risk* for developing Alzheimer's, but they are not deterministic. (If you're considering genetic testing, keep that crucial difference in mind. I recommend that you enlist the help of a genetic counselor to interpret your results, because they can be confusing.) There is also evidence that those with a genetic risk of Alzheimer's (that is not due to the rare deterministic genes) can nonetheless lower their risk of developing the disease. According to one study, people with highest genetic risk who maintained the brain-healthy lifestyles we cover in this book lowered their risk of dementia by half, compared to people of similar genetic risk who did not maintain them.[12] Studies have also investigated how the ApoE4 gene (a variant of APOE known to raise the risk of Alzheimer's) influences the risk of Alzheimer's worldwide. For example, intriguing research found that those of West African ancestry who carry the ApoE4 gene and live in Nigeria have a lower risk of developing Alzheimer's than those of West African ancestry who carry the ApoE4 gene *and live in the United States*.[13] These findings show that genes aren't everything; lifestyle and environment are also involved and impact ApoE4's role in developing Alzheimer's.

Men, Women, and Alzheimer's

We hear that men have a better sense of direction but can't remember what to get in the grocery store. Women are said to be better at multitasking but worse at parallel parking. These often-cited differences in the male and female brains have helped sell millions of books and provided endless hours of debate at cocktail parties. The truth is science has proved these differences in the male and female brains to be myths.[14]

But what *is* true, and often not discussed, is that women account for two-thirds of all cases of Alzheimer's disease. In the past, it was assumed that women suffered from Alzheimer's at a greater rate because they lived longer, but now we know that's not the whole picture. As we have discussed, there are several key underlying factors to the roots of dementia. Unfortunately, some of these underlying issues are either underdiagnosed or misdiagnosed in women.

For example, sleep apnea is one of the most significant risk factors for memory loss. A study uncovered that those with untreated sleep apnea lose their memory on average ten years before the general population.[15] In women, sleep apnea symptoms are often similar to menopause. A study found that 80 percent of sleep apnea cases in women are underdiagnosed. Additionally, since the symptoms of many of the root causes of dementia, such as sleep apnea, cardiovascular health, anxiety, and depression, are different in women than men, they're often not treated effectively.[16] The misdiagnosis and undertreatment of these underlying conditions are two key factors in the larger number of women suffering from Alzheimer's.

Women have unique risk factors for Alzheimer's as compared to men. For example, the reduction in estrogen levels during menopause can shrink the brain by decreasing the volume of gray matter in the brain. A 2021 study found that increased cumulative exposure to estrogen (throughout one's life span) may protect against this shrinking.[17] Increased exposure to estrogen can come from hormone therapy during menopause or from having children earlier in life. This evolving, critical, and individualized area of women's health is important to discuss with one's primary care provider.

Both women and men can improve brain health, including scores on cognitive tests, by taking preventative measures (like those covered in this book) along with personalized interventions based on underlying conditions.[18] In fact, a 2022 study found that women who follow these types of preventative strategies might even obtain a greater brain benefit as they improved their scores on cognitive tests more than men improved after eighteen months of the brain health intervention.[19]

Lowering Risk

Because in most cases dementia takes a long time to progress, we have time to act. We can improve or even arrest developing symptoms.

We have the most information on how to lower the risk of Alzheimer's and vascular dementia, unlike Lewy bodies and Parkinson's disease, whose root causes are unknown. However, as evidence emerges, the lifestyle factors that are shown to reduce the risk of Alzheimer's and vascular dementia are being explored as possible avenues for reducing risk in Parkinson's and Lewy body dementia.

There are medications that can delay the progression of symptoms. However, medications that solely try to eliminate plaques and tangles have been disappointing, as they haven't clearly or consistently reversed disease progression in multiple clinical trials.[20] Why? Recall that in many cases Alzheimer's is not a disease with a single cause. There are multiple factors. But in fact, this is hopeful news.

We can borrow a page from oncologists, who have a theory called the multi-hit theory. It describes the idea that there's not one factor that causes a cell to become cancerous; multiple things have to go awry before that happens. The same theory applies to brain health: It is not just one risk factor that leads to brain aging, dementia, or Alzheimer's in most cases; it is the accumulation over time.[21] However, like the straw that broke the proverbial camel's back, there may be a triggering factor that causes a healthy cell to become a diseased one. By looking at the risk factors for dementia and Alzheimer's (such as diabetes, heart disease, inflammation, and lack of sleep), we can take as many straws off of the camel's back as possible. We can leverage what we *do* have control over to lessen our risk and create a resilient brain.

Diabetes is one of the major risk factors—so that's what we will look at next.

CHAPTER 7

Insulin Resistance and Diabetes

DO YOU HAVE A SWEET TOOTH? HUMANS DON'T CRAVE SUGAR just because it's delicious—seeking out sugar is programmed in our DNA. In fact, DNA is partly made of sugar.[1] At your essence you are partly sugar, and blood sugar is the primary fuel of the brain. Not enough of that fuel and you have no energy; too much and you can destroy blood vessels and tissue, leading to premature aging and wrinkles (due to collagen breakdown in the skin) and cardiovascular disease (due to damage to the heart muscle).[2] And, of course, too much sugar also does increase our risk of diabetes.

Simply put, excess sugar is poison to the brain—and a large part of that is because excess sugar can lead to insulin resistance, prediabetes, and diabetes, which are an often overlooked fast track to brain issues and accelerated brain aging. We'll define all of these terms in a minute, but to underscore why this is so important, consider:

- A 2021 study found that insulin resistance doubles the risk of developing major depressive disorder, even if someone has never experienced depression previously.[3]

- Untreated diabetes raises the risk of developing Alzheimer's by 65 percent, making it the greatest risk factor aside from age.[4]

To define and discuss the critical aspects of brain health and aging related to insulin resistance, prediabetes, and diabetes, let's do a quick crash course on the key players—insulin and sugar—and how they work. To do that let's think about Will Ferrell and Chris Kattan in the movie *Night at the Roxbury*. (Did you just start hearing the song "What Is Love" in your head?) Will and Chris play two characters who do everything they can to get past the red velvet rope and into an exclusive club. Think of Will and Chris's characters as sugar, and your cells as the club they want to get into.

After we eat food, the sugar from the meal gets into the blood and goes cruising around. Sugar needs to get inside cells, where it is used as fuel. But sugar can't get into cells on its own. Of course your cells want sugar; they need it survive. But the ability of sugar to get into cells has to be tightly regulated. Not too much sugar and not too little. So you can think of insulin as being the cool friend on the VIP list who knows the secret knock to gain access to the club, while sugar is the +1 who can't get in alone. Insulin attaches to cells, gives the secret knock, and then a little tunnel opens up outside the cell and allows the sugar into the cell. Essentially, insulin is a hormone that instructs our cells to process the sugar that is in our blood after eating.

Without any or enough insulin, sugar remains in the blood, where it destroys blood vessels and leads to damage to the heart, kidneys, and brain, as well as loss of vision and limbs. We want sugar to be in the cells of our muscles and organs, not in our blood. Too much sugar left in the blood is toxic.

So how do we describe what happens when there's too much sugar in your blood?

Insulin resistance is what happens when the cells in the body are no longer listening or responding to insulin—even though the pancreas is still able to secrete insulin into the bloodstream. Think of this as if the insulin is doing the secret knock on the door of the cells, but the door to let the sugar in is not being opened. Thus, sugar levels build up in the blood instead of getting into the cells. Insulin resistance can be caused by a variety of factors, including lack of physical activity, too much sugar in the diet, unmanaged stress, and not getting proper sleep. One in three people in the United States has insulin resistance.[5] Many are not aware they have this condition.

Prediabetes is insulin resistance that precedes full diabetes. It happens when the pancreas is working double-time to cover the loss of insulin functionality. One of the most important things about prediabetes is that it's completely reversible—but if it's not treated, it can be dangerous.[6] The problem with treatment, though, is that 90 percent of those with prediabetes are not aware they have the condition because there are often no prediabetes symptoms. It probably goes without saying that it's impossible to treat something you don't know you have. However, there are some symptoms that are often overlooked because they're so subtle, such as tiredness or acetone-flavored breath.[7] (This is why it is important to track sugar levels at your doctor's visit with a blood test we will include on the "One Important Sheet of Paper.")

A 2021 study found that those with prediabetes were 42 percent more likely to have cognitive decline over the following four years than those without. What's more, those with prediabetes were also 54 percent more likely to develop vascular dementia within the next eight years.[8] Those with prediabetes have an increased risk of suffering from heart disease and heart failure—even before progressing to diabetes.[9]

Diabetes is what happens when your body's ability to make or respond to insulin doesn't work correctly. Going back to our VIP club, this time, think of insulin as that cool friend with all the connections—but this friend is flakey and doesn't show up. Thus sugar cannot get into the cells (club) on its own. There are two types of diabetes, called (appropriately enough) type 1 and type 2. Type 1 diabetes is thought to be an autoimmune disease in which the body destroys the cells in the pancreas that create insulin. It is treatable but requires lifelong management.

The vast majority of diabetes cases—90 to 95 percent—are type 2 diabetes. In this condition, the body doesn't use insulin properly. Sensing that the body's normal response to insulin has been blunted, the pancreas releases *more* insulin than normal to allow cells to open up and let the sugar inside. This would be like sugar trying to find more people who know the secret knock to get into the exclusive club. This is effective in the short term, but eventually, the cells stop responding again and blood sugar rises. If blood sugar levels stay elevated chronically, this leads to type 2 diabetes. That level of insulin resistance and the inability to use sugar as fuel in the brain can also lead to neurological disorders, mood issues, and problems with memory, focus, and other aspects of cognitive ability. This is particularly alarming when you realize that one in two Americans has type 2 diabetes or is prediabetic.[10]

Changes to diet, lifestyle, and, in some cases, medication can treat (and sometimes reverse) insulin resistance and type 2 diabetes.

Unfortunately, in many ways the modern world seems to set us up to develop these conditions. Inexpensive and easy-to-obtain food is often loaded with added sugar and high fructose corn syrup. Food is available 24/7 via drive-throughs and delivery without much physical effort. Our lifestyle can easily be sedentary, which affects not just our physical health but also our brain health, as a sedentary lifestyle can make it harder to sleep, lessen blood flow to the brain, and increase inflammation, which can damage insulin production. Just because our modern world is set up to increase the likelihood of these conditions doesn't mean that it's inevitable.

There are multiple consequences to not treating diabetes, both type 1 and type 2. For instance, it can lead to damage to the heart, pain, kidney disease, loss of limbs, and brain aging. Studies published in *Movement Disorders* found that type 2 diabetes is also associated with an increased risk for, and faster disease progression of, Parkinson's disease.[11] As mentioned, untreated diabetes raises the risk of developing Alzheimer's by 65 percent, making it the greatest risk factor aside from age. The links between diabetes and neurological conditions are so strong that treatment and prevention strategies for type 2 diabetes may be repurposed to treat Alzheimer's and Parkinson's.[12] In fact, the connection between Alzheimer's and diabetes is so profound that some brain researchers refer to Alzheimer's as type 3 diabetes.[13]

What Is the Diabetes-Alzheimer's Link?

In Alzheimer's, the insulin resistance that affects the body also takes place in the brain: Brain cells can no longer use sugar as fuel, leading to brain and memory dysfunction. This is very much like a car that runs out of gas. This shared underlying mechanism is why Alzheimer's is referred to as type 3 diabetes—though, of course, Alzheimer's is a disease that occurs because of an accumulation of multiple factors. While those multiple factors include plaques and tangles, as we mentioned earlier, they also include brain metabolism. Brain metabolism is the ability of the brain to use sugar as fuel for each of your brain cells. You need that fuel to think, focus, remember, and make decisions.

Earlier in this book, I suggested thinking of your brain like your desk—it's harder to get things done when your desk is messy. That messy desk analogy can provide some insight into how diabetes and Alzheimer's are linked. Your body is fantastic at recycling. It recycles and rebuilds parts of you every day—your body is a complex set of proteins that are constantly being taken apart and reassembled. Enzymes act like scissors, chopping up proteins into small pieces to be rebuilt and reused. One of those proteins is insulin. The enzyme that breaks it down—insulin-degrading enzyme—turns out to do two things: it chops up both insulin and amyloid plaques in the brain.

Now imagine you have two tasks: a cluttered desk hiding a printout you need for work and taxes that are due in a few hours. What happens? If you choose to focus on your taxes, you won't find that printout for work. If you look for the printout, you'll be paying a penalty to the IRS. When an individual develops insulin resistance, the body overproduces insulin. This is called the compensatory hyper insulin response. The insulin-degrading enzyme is so busy keeping up with the excess insulin that it cannot keep up with disposing of the amyloid plaques. The brain trash builds up, like the clutter on a messy desk, making it harder to do what needs to be done.

This enzyme is just one link between Alzheimer's and diabetes. Excess blood sugar itself is toxic, and it damages brain cells.

What Leads to Diabetes?

Now that we know diabetes can lead to major problems in brain health, you may be wondering if you're at risk. Let's take a look at what causes insulin resistance and diabetes in the first place.

Our Genes

Diabetes is a complex mix of genetics and environment. For type 2, heritability ranges from 20 to 80 percent, based on a variety of factors, including whether or not one or both parents have the condition. The heritability of type 1 diabetes is complex as well, with multiple variables. For example, a woman with type 1 diabetes has a 1 in 25 chance of passing on the condition to her child if she has the child before the age of twenty-five. If the child is born after the mother is twenty-five, the risk drops to 1 in 100. A man with type 1 diabetes has a 1 in 17 probability of passing on the condition.[14] What does that tell us? Since genes alone do not cause diabetes, our environment and lifestyle play a critical role as well.[15]

What and When We Eat

Perhaps it won't shock you to hear that our diets play a role in the development of insulin resistance and diabetes. In particular, you may have heard that eating a lot of sugar causes diabetes. As I mentioned at the beginning of the chapter, our bodies do need *some* sugar to function. But excess sugar turns into fat in the body: an extra 500 grams of sugar a week can translate to gaining an extra pound a week.[16] Excess fat is a risk factor for developing diabetes and inflammation. Fat cells release inflammatory factors. A 2020 study found that obesity is linked to a sixfold increased risk of developing type 2 diabetes.[17] This factor was more significant than genetics and lifestyle alone.

It's important to note that natural sugars found in foods like fruit and unsweetened dairy products are not the enemy. The enemy is *added* sugar—namely, the sugars added to processed foods.[18] We'll get into this in detail in chapter fourteen, but for now, know that as long as the sugar you consume is from natural sources like whole fruit and unsweetened dairy, you're probably doing OK.

It's Personal

Which food increases blood sugar more, sushi or ice cream? The answer seems like an obvious one; it must be ice cream, right? Well . . . not quite. A study published in *Cell* tracked eight hundred people, recording the various foods they ate and measuring their blood sugar. The researchers found that the foods that raised blood sugar were highly individualized.[19] Thus, it was ice cream in some people, and in other people, it was sushi that raised blood sugar. Why the difference? Gut bacteria. (Remember, that's the answer to a lot of questions!) A healthy gut is a crucial step in lowering the risk for diabetes and maintaining a healthy, youthful brain. See chapter fourteen for more information on maintaining a healthy gut microbiome.

Interestingly, as much as *what* we eat, *when* we eat may play a role in our risk for developing diabetes. A study presented at the Endocrine Society's annual meeting in 2021 found that people who started eating before 8:30 AM had less insulin resistance and lower blood sugar levels than those whose first meal of the day was later.[20] Another indication of the importance of time restricted eating comes from a 2021 analysis in *Endocrine Reviews*, which found that limiting when we eat to a consistent, less-than-twelve-hour window of time during the day can in some cases improve metabolic health.[21]

The effects of intermittent fasting are less clear. Intermittent fasting refers to cycles of voluntary fasting and eating throughout the day. For example, one intermittent fasting protocol has individuals consume all their meals within an eight-hour period during the day, and refrain from eating for the other sixteen hours. Some studies have shown intermittent fasting helps manage and lower the risk of diabetes.[22] However, other research presented at the European Society of Endocrinology found that fasting every other day might increase diabetes risk by deregulating insulin.[23] Any form of fasting should be discussed with a personal physician, as it can be dangerous to alter calorie intake and eating schedules drastically.[24]

What is *never* good for your blood sugar is eating food with added sugar right before you sleep or in the middle of the night. Our brain clock, or suprachiasmatic nucleus, plays a role in metabolizing food. Food is more effectively metabolized during waking hours. Thus, it is possible that eating at times when metabolism is not as effective could lead to leftover sugar in the blood, which can raise the risk of insulin resistance and brain aging.

Pollution and Toxins

Pollutants in our environment disrupt our insulin function and are an under-reported risk factor for diabetes. A study reported that air pollution accounted for 3.2 million, or *14 percent,* of all new diabetes cases globally.[25] Marginalized communities of lower socioeconomic status often live near areas of high pollution.[26] Furthermore, there is an increased prevalence of type 2 diabetes in mid-life for those of lower socioeconomic status.[27] Pollution can be one factor in this increased risk.[28] Chapter fifteen in this book is dedicated to minimizing contact with specific pollutants and toxins.

Disrupted Sleep

Sleep is critical to lowering your risk of developing diabetes. If your brain clock is thrown off, it can negatively impact the release of insulin. A study done at the University of Chicago Medical Center linked insomnia to high insulin resistance in people with diabetes.[29]

Your One Important Sheet of Paper: Test Blood Sugar

Testing blood sugar is another question we will add to that "one sheet of paper" to bring to your next physical. The only way to diagnose prediabetes or diabetes is using blood tests such as the hemoglobin A1C blood test, which tests for the percentage of hemoglobin proteins in your red blood cells that are coated with sugar. According to the Centers for Disease Control and Prevention, "An A1C below 5.7% is normal, between 5.7 and 6.4% indicates you have prediabetes, and 6.5% or higher indicates you have diabetes."[30] Other common tests for diabetes and prediabetes are the glucose tolerance test (where your blood sugar is measured before and after you drink a sugary beverage) and the fasting blood sugar test (where your blood sugar is measured after a night of not eating). Your health-care provider will work with you to understand your results and suggest a course of action to get your blood sugar into normal ranges or keep it there.

Treating Insulin Resistance, Prediabetes, or Diabetes

If you are diagnosed with diabetes or prediabetes, don't despair. The condition can be effectively managed and even reversed through medical intervention and lifestyle changes. What's more, there's hope for even those with severe diabetes (which is good news for brain health). For some individuals who have diabetes and are severely overweight, surgery may be a viable treatment option. Research found that 37.5 percent of patients with severe diabetes who received metabolic surgery such as gastric bypass were diabetes-free *ten* years after the procedure. Surgery was more effective than medication and lifestyle interventions in these cases.[31]

For less severe diabetes and prediabetes, though, medicine can be incredibly effective. For instance, the drug metformin limits the amount of sugar released into the blood and helps muscle cells "hear" the insulin knocking more effectively. Studies found that prediabetic individuals placed on metformin lower their risk of developing diabetes by 30 percent. Most importantly for our discussion of brain health, a six-year study published in *Diabetes Care* found that older patients taking metformin for type 2 diabetes had a similar rate of decline in cognitive function as those without diabetes who were not taking metformin.[32] Another study found that both metformin and a Mediterranean diet are good candidates for protecting brain health for those with diabetes, but adherence to a Mediterranean diet might outweigh the brain-protecting effects of metformin for those with diabetes.[33] Other studies in the *Annals of Neurology* uncovered that some treatments with the diabetes medication pioglitazone significantly decreased the risk of dementia—sometimes to the point that people with diabetes developed dementia *less* often than non-diabetics. Specifically, the risk of developing dementia was around 47 percent *lower than in non-diabetics.*[34]

With or without medication, small lifestyle tweaks can have a big impact on your brain health:

- Minimize added sugar.
- Move for fifteen minutes after each meal (we'll see why in chapter twelve).
- Consider eating breakfast before 8:30 AM.
- Try not to eat in the middle of the night.

Part three of this book will explore what you can do in more detail. For now, we can take heart. By effectively managing insulin resistance and treating diabetes, we can protect the brain and significantly lower our risk for a long list of health conditions, including mood disorders such as depression, as well as Alzheimer's and dementia.

CHAPTER 8
Inflammation and the Brain

WE'VE TALKED ABOUT HOW A BRAIN THAT FILLS UP WITH TRASH can lead to aging. The brain has three ways to naturally take out the trash: one, learning new things; two, sleep; and three, the action of microglia (the "scavenger" immune cells I mentioned in chapter two). But sometimes, the microglia can get confused, and instead of eating trash, plaques, tangles, and toxins, these cells start to gobble up healthy brain cells and turn them into (more) useless brain trash. So what confuses the microglia? Chemicals produced by chronic inflammation. When the immune system mistakenly attacks an otherwise healthy part of our body, it creates inflammation, sparking a vicious cycle. We are learning that brain diseases like Alzheimer's, anxiety, and depression can be rooted in the immune system incorrectly attacking the brain.

The last thing I want to do is scare you into panicking about your brain health every time you get a scratch or a cold, worried that the inflammation that accompanies the healing process is causing your immune system to go haywire. That's why, before we go any further, I want to take time to note the difference between acute and chronic inflammation. Acute inflammation is sudden inflammation that can be caused by an infection or injury. Essentially, *acute* means that there *is* inflammation, but then it goes away as you heal. We are not concerned about inflammation caused by a scratch or bruise or infection that heals—the inflammation goes away as you get better, and your body knows how to handle that. It is chronic inflammation, which is persistent, that we are most concerned about (except in instances of acute injury to the brain—but we'll get to that in a few pages).

Studies done at the Max Planck Institute for Biology of Ageing found that increased chronic inflammation causes the aging process to accelerate.[1] Think of inflammation like a fire raging as it attacks and damages different parts of the body. Inflammation can impair insulin-releasing cells, which can cause type 2 diabetes; it can harm the heart and lead to cardiovascular disease; it can also increase the risk of cancer by damaging DNA. While the risk of developing these diseases increases with age and these diseases themselves potentially accelerate aging, chronic inflammation can put these processes in hyperdrive. Additionally, there is evidence that inflammation can also shrink the hippocampus, disrupting memory and learning.[2]

Dental Health Is Brain Health

A study done at the University of Bergen uncovered a surprising link between gingivitis and a higher risk for Alzheimer's.[3] Inflammation in the gums can release factors into the bloodstream that confuse the microglia in the brain. To keep our microglia focused on just eating brain trash and not healthy nerve cells, we need to brush our teeth, floss, and have regular dental checkups.

The speed at which you age is related to the ends of your chromosomes, tightly coiled DNA found in the nucleus of each cell. You know that your DNA is the precious plan that dictates how you're built from head to toe. Think of your DNA as a blueprint—or even as that ubiquitous double helix—and right now, think of chromosomes as shoelaces. If you do a lot of crossword puzzles, or were watching the Disney Channel in the 2010s, you might know that the little plastic things at the ends of laces are called aglets, which protect your laces from unraveling (and if you didn't know before, you've just learned a new thing and helped protect your brain). The ends of your chromosomes feature telomeres, something very much like an aglet.[4] These caps protect your chromosomes just like aglets protect your shoelaces. But as we age, these protective caps get shorter and shorter—in fact, essentially, our true biological age can be correlated with how much protection we have at the end of our chromosomes.[5] If the protection goes away, the precious DNA blueprint becomes corrupted and the body may fail to make the proper immune cells, brain cells, heart cells—and so on. Lifestyle can affect telomere length, so that people of the same chronological age can have telomeres of different lengths, and thus different biological, or true, age.

Inflammation causes the telomeres to shorten, almost as if you set fire to your aglets.[6] Inflammation can damage telomeres—and damaged telomeres can increase inflammation, perpetuating a vicious, incendiary cycle that ages you. Even low levels of chronic inflammation can be dangerous. They are like a small fire in your home that over time hits a main pipe or key electrical panel—even if the whole building isn't burned to the ground, there's still extensive damage. And, similarly to how a fire spreads, often inflammation can start somewhere else in the body, such as the gut, and travel to the brain.

When certain autoimmune disorders are not treated effectively, the resultant inflammation leads to a significantly increased risk of Alzheimer's and dementia.[7] Even without an underlying condition, aging and inflammation are so closely linked that there's even a term for the type of chronic inflammation that accompanies growing older: inflamm-aging. There is also strong evidence that inflammation plays a role in depression, fatigue, mood, and anxiety disorders, all things that also age your brain, as we'll see in the next chapter.[8] Thus, to protect our DNA and keep our brains and bodies young, we must put out the fire of chronic inflammation. (By the way, the lifestyle factors that protect your telomeres are the same factors that age-proof your brain, which we discuss in part three of this book.)

A Quick Refresher on Your Immune System

In chapter two, we discussed how your immune system is basically a complex army of killer cells that attack and kill off dangerous pathogens, and peacekeeper cells that calm those killer cells down. There is a delicate balance between these two types of immune cells. As we age, the balance can shift, giving us more active killer cells and fewer peacekeepers. This imbalance is why older adults tend to heal slower than younger adults and children—the peacekeepers are not doing their job, there's no longer enough of them to keep up with killer cells, and the killer cells keep attacking. Autoimmune diseases and chronic inflammation are, despite all their complexity, essentially killer cells attacking one's own body. Obesity, heart disease, and diabetes can either be caused or partially caused by immune dysfunction. These conditions and illnesses can also throw off the immune system's effectiveness and create inflammation, leading to yet another vicious cycle.

Your Important One Sheet of Paper: Test for Inflammation

Make sure that a blood test to measure inflammation becomes part of your annual physical. Consider tests such as a CRP (C-reactive protein) blood test. (An elevated CRP has been linked to an increased risk of dementia.) This test is not automatically included in standard blood work, but you can ask your doctor to include this as part of your annual exam. If your CRP is high, you can take the steps discussed later to lower it.

The Mind-Mood-Immune Connection

If you've ever been in a locker room before a big game, you will observe athletes pumping themselves up, often with the encouragement of their coach. They might even look like they are angry. Anger, if used correctly, can be motivating. In this context, anger can be a positive force as long as it is managed, not violent, and is used to make improvements. A study published in *Aggressive Behavior* investigated rugby players who used anger to get inspired before a big match.[9] The researchers checked the players' blood during these bouts of pre-game anger and found an acute surge in the killer, pro-inflammatory cytokines. Cytokines are like the security guards of your immune system. A surge can have some health benefits, such as killing off any dangerous viruses or bacteria. This surge is some of the evidence that points to the connection between mood and the immune system.[10]

Negative and positive mood states can influence the immune system, and vice versa. For instance, increased levels of certain types of pro-inflammatory cytokines can increase the risk of mood disorders and depression.[11] On the other side of the coin, positive mood states are typically associated with a balanced immune system—but as the rugby study shows, even so-called negative moods can be good for health at times.

We have our ancestors to thank for this mind-mood-immune connection. Take a moment and imagine you are with your prehistoric ancestors. You're out hunting for food and see a hungry bear heading your way. You're going to have to fight that bear, and, even if you win (which . . . is unlikely), you *are* likely to be injured. So, your body needs to prepare to manage those injuries. When we're stressed or angry, the body releases the hormone cortisol, which increases our heart rate and blood pressure, priming us to fight or flee. Stress also triggers the release of pro-inflammatory cytokines, which gave our ancestors a head start to fight off any dangerous infections from an injury sustained during a fight. Thus, there was a survival advantage to having certain moods trigger the immune response (and those who had that connection had a better chance of living to pass it on). After the threat went away, the prefrontal cortex would release calming hormones to counteract the effects of cortisol and get our ancestors back to baseline.

Our bodies still respond to threats by releasing cortisol and thus cytokines. The problem is that instead of being threatened occasionally by a saber-toothed tiger or spear-wielding enemy, we're stressed daily. A study published in *Brain, Behavior, and Immunity* found that when subjects were told to *remember* something that made them angry or stressed, it could produce a surge in killer T cells.[12] But while cortisol is beneficial in small, sporadic doses, when too much cortisol is released, or it is released too often, it can cause an increase in inflammatory immune cells and a decrease in the anti-inflammatory cells, leading to chronic inflammation and all of the damage that entails: a reduced ability to fight disease and harm to the hippocampus and organs, which can raise the risk of memory loss, anxiety, mood changes, depression, and dementia. Even low levels of stress, if constant, can lead to cortisol and killer cells being released in the bloodstream, leading to a chronic inflammatory response and throwing off the balance of the immune system.

Good News for Dog Lovers

A study published in *Academic Emergency Medicine* asked if something could be done to bring high stress levels down in emergency room physicians and nurses.[13] The researchers had the doctors and nurses spend five minutes playing with and petting a therapy dog. After five minutes, researchers tested participants' blood and saliva for levels of pro- and anti-inflammatory cytokines and the stress hormone cortisol. They found workers had returned to a more balanced state, leading to less inflammation and an improved overall immune function. If you're wondering if it's just dogs that provide benefit, a 2022 study found that pets of any kind can help improve memory.[14] The authors of the study stated that part of this memory boost is due to the antistress and anti-inflammatory benefits of pets. Those of us who own pets often feel their presence as soothing, and it turns out there are real physiological underpinnings to the happiness we feel when we're around our furry (or feathered) friends.

If you are not a pet owner, having social support has also been shown to help balance the immune system.[15] We'll explore that further in chapter thirteen.

Brain Injury and Inflammation

Earlier in this chapter, I said that when it comes to brain health, chronic inflammation is more concerning than acute inflammation. But there is some acute inflammation we need to worry about—specifically, that which results from severe head injuries to the brain.

In the United States alone, over 23 million adults over the age of forty have had a head injury that resulted in a loss of consciousness.[16] But a head injury can cause lasting damage even if you don't lose consciousness. Multiple studies have uncovered a link between head injuries, such as concussions, and dementia.[17] Even a mild brain injury can have long-term negative impacts on brain health decades later: a 2021 study done at the University of Pennsylvania found that just one single head injury can increase a person's risk of dementia by 1.25 times compared to those without head injuries. It also found that nearly one in ten dementia cases can be attributed to at least one prior head trauma or injury.[18]

Studies have uncovered that each successive head injury increases the risk of dementia.[19] But as dangerous as a brain injury can be, it's a risk factor that we can often minimize by simply using preventive measures, such as wearing seat belts and helmets.

Brain injuries, including concussions, raise the risk of dementia for two key reasons. The first is that when the brain swells, it disrupts the washing process that removes toxins from the brain.[20] The second is that the inflammation that occurs after a brain injury can cause the microglia to attack healthy brain cells instead of cleaning waste. The lifestyle factors covered in this book that optimize the washing process and suppress inflammation are essential parts of comprehensive treatment for traumatic brain injuries.

A Final Surprising Connection to Give You a Smile

A small but intriguing study done at Loma Linda University looked at twenty people with diabetes, hypertension, high blood pressure, or heart conditions.[21] Those in the study were put on the same medications based on their specific conditions and were followed for a year. The researchers then divided them into two groups. One group got to do something the other didn't: they watched comedies for thirty minutes a day.

The researchers tracked both groups and compared their blood for stress hormones, cholesterol, and inflammatory cytokines. Month after month, the group that watched the comedies had lower stress hormones, and their good cholesterol went up by 26 percent. Inflammatory cytokines decreased by 66 percent in the group that watched comedies, versus 26 percent in the control group. This may be a small study, but it's something to consider regarding how our day-to-day mood is tied to our immune system and multiple aspects of our health. Remember to schedule some time in your day to have some fun and laugh.

This chapter's discussions of mood and brain health may have made you wonder about how mental health affects the brain more generally. That's what we'll cover in the next chapter.

CHAPTER 9

Mental Health Is Brain Health

STUDIES COME ALONG FROM TIME TO TIME THAT CAUSE SCIENTISTS to pause and do a double-take. This is one of them: Researchers spent thirty years following 1.7 million people to see if there was a link between developing dementia and mental health disorders such as depression, anxiety, and bipolar disorder.[1] When the study was published in 2022, it showed that on average, onset of dementia occurred five years earlier in those who had a mental disorder than those who didn't have a mental disorder. The researchers also analyzed chronic diseases such as cardiovascular disease, stroke, diabetes, gout, chronic obstructive pulmonary disease, traumatic brain injury, and cancer, and were surprised that the risk of developing dementia was more strongly associated with mental disorders than with chronic physical disease.

It is important to note that each of these conditions can raise the risk of brain aging and dementia. In the 2022 study, this link between mental health disorders and dementia was present in males and females and applied to Alzheimer's and non-Alzheimer's dementia. So while we know that mental health is critical for our day-to-day wellness, we now understand it can profoundly impact brain aging and dementia years down the road. Understandably, just hearing that news can be anxiety provoking. To be clear, having a mental health disorder does not mean someone is destined to develop dementia. But there is an elevated risk, and this understanding opens new avenues to protect the brain.[2] To reassure you, we'll get to those—the same protective strategies I'll outline in part three are also key lifestyle interventions for treating mental health disorders. Before we get there, though, let's understand the science of mood, mental health, and brain health.

Mental health disorders are connected to dysfunction of mood, but they're not the same (put more simply, a bad mood is not a mental health disorder). Mood tells us how good versus bad things are in our internal and external world—it's a key to our survival and our reproduction.

The Brain Science of Moods

Even though mood is a temporary state of mind, it plays a key role in connecting our immune system, hormones, nervous system, and brain. However, when mood becomes something one absolutely cannot control, it becomes a medical issue. It can have a devastating impact on brain health and prematurely age the brain.

What exactly is mood? Imagine a typical day: You're driving to work, and someone cuts you off. You go to the grocery store to pick up some milk, and somebody ahead of you in the express checkout has too many items. You go to a movie, and somebody is talking on their cell phone. If you're like me, you probably feel your mood dipping just thinking about those things. But while you might think just one of those incidents could put you in a prolonged bad mood, the truth is, it probably couldn't. If the only thing that happened was someone cutting you off, you might feel annoyed—but it shouldn't ruin your entire day. That's why moods—especially bad ones—are more often the products of an *accumulation* of moments.

An interesting study put people alone in a room with a bunch of doughnuts and told them not to eat the treats for twenty minutes.[3] Some people ate the doughnuts anyway. Some people abstained. (I don't know how, but they did!) Once the twenty minutes were up, someone would walk into the room and insult the study participant. Who do you think was in a worse mood after being insulted—the doughnut eaters or the abstainers?

The people who resisted eating the doughnuts were more likely to be in a bad mood and more likely to get aggressive. That's because most people want to eat the doughnuts. That desire activates the limbic system (which is made up of your amygdala, hippocampus, and hypothalamus)—the "want" part of the brain. When you *resist* the things you want, you engage another part of your brain, the prefrontal cortex, which we discussed in chapter five; it acts as a brake and sends signals to suppress the want. Remember, the prefrontal cortex is like a cell phone battery, charging overnight as we sleep. As the day goes on and we're coping with the demands of daily life, it starts to deplete, and so do focus, willpower, and mood.

The people who didn't eat the doughnuts were using willpower, which drained the prefrontal cortex. Thus, when they were insulted, they didn't have the energy left to control their moods. (This is also why people are most likely to go off their diets at nighttime.)

Your mood is also connected to your health in surprising ways. For example, mood can impact how successful a medical procedure is. A study assessed people's moods before having an operation and found that people who went into a procedure in a bad mood—who were feeling fear, anger, or sadness—had significantly more negative effects from the surgery.[4] Twenty-two percent of the people in bad mood states had a hard time recovering, while only 12 percent of people in better moods struggled to bounce back. It didn't matter how good the subject's mood was—just that they were not in a bad one. You don't need to tap dance into a hospital visit, but it is important to not be in a negative mood state even during an understandably stressful time. This relationship between mood and healing is not very surprising when we think about how our mood impacts our stress levels. Having too much stress or cortisol can increase inflammation. Too much inflammation, and the body doesn't heal. Our mood, our healing, and our overall heath are all connected.

Mental Health Disorders and Brain Aging

Mental disorders like depression, bipolar disorder, and anxiety are exceedingly complex—and of course, they are not the only such disorders. For the scope of this book, I am only discussing aspects of these three common conditions to place them in context of brain aging.

Three Common Mental Health Disorders

Anxiety

Anxiety and fear are normal emotions. (The ability to worry and experience fear is important to our survival: humans are naturally afraid of snakes, spiders, heights, animals larger than we are with sharp teeth and claws, and people sneezing around us.[5] If your ancestors hadn't been scared of these things, they probably wouldn't have survived. We are the offspring of the scared.) We feel fear in the presence of an immediate threat. Anxiety is a feeling of worry or nervousness about an event in the future or a hypothetical event. Fear is a response to something happening in the moment. Anxiety is anticipation. Both can provoke physical responses—fear can manifest in fight-or-flight response, while anxiety can manifest in tension, worry, and physical symptoms such as increased blood pressure, sweating, dizziness, rapid heartbeat, or trembling.

However, when anxiety negatively impacts our ability to get through the day, is out of proportion to what's causing it, or persists after a stressor is gone, it becomes a disorder. Fear becomes a disorder, such as a phobia, which is an irrational fear of something that poses no danger.

Anxiety disorders, which also include generalized anxiety disorder, panic disorder, OCD, PTSD, and social anxiety disorder are the most common psychiatric illnesses among children and adults. Forty million US adults, or 18 percent of the adult population, suffer from an anxiety disorder.[6]

While many types of anxiety are highly treatable and symptoms can be managed, it is estimated that only a third of people with anxiety disorders get treatment.[7] It's important to take care of anxiety because an anxiety disorder increases the risk of having a heart attack or stroke. Somebody with heart disease and an anxiety disorder has double the risk of a heart attack.[8] Anxiety also impacts the immune system, as excessive release of stress hormone can lead to inflammation.[9]

Depression

Depression is a medical illness. Before the Covid pandemic it impacted about 8.5 percent of the population in the United States. Research from Boston University found that in 2020, that rate jumped to 27.8 percent—and, in 2021, climbed to almost *33 percent of the US population.* Symptoms vary from person to person, but two significant signs of depression are losing interest in something that used to bring enjoyment and having feelings of sadness or hopelessness that last for two weeks or longer.[10] Depression manifests in many different ways because it disrupts multiple parts of the brain, as we'll see in the next section.

Bipolar Disorder

Bipolar disorder (previously known as manic depression) is complex, but a simple description is that it causes extreme fluctuations in mood, from very high, called mania, where one feels out-of-control high energy and excitement, to poor, or a state of depression. Bipolar disorder is, at its essence, the inability to regulate mood based upon the environment—that is, aspects of an individual's life do not necessarily match their mood. There are two types of bipolar disorders: bipolar I and bipolar II. Bipolar I disorder is described as severe mania and depression. Bipolar II disorder involves depression with less severe mania.[11] In the United States, 1 to 2.6 percent of the population suffers from either bipolar I or II.

Anxiety, depression, and bipolar disorder can all lead to exhausted brain cells and chemical imbalances, which can prematurely age the brain and raise the risk of dementia. A 2021 study found that occurrence of depression in early adulthood (approximately eighteen to forty years old) increases the risk of dementia in older age by 73 percent; those with depression in early adulthood are more likely to have lower cognition ten years later.[12] For those who had depressive symptoms later in life, the risk for dementia increased by 43 percent compared to those without depression. Studies suggest that anxiety disorders increase the risk of dementia by 29 percent, and bipolar disorder increases the risk of dementia by nearly three times compared to those without bipolar disorder.[13]

The Potential of Lithium

Lithium is often used to treat bipolar disorder. A 2020 study found subjects with bipolar disorder receiving lithium as a treatment had approximately half the risk of developing dementia as those with bipolar disorder who were not taking lithium.[14] Larger studies are needed, but the role of lithium in reducing the risk of dementia in the general population is a fascinating and emerging area of study.[15]

Why would depression, anxiety, and bipolar disorder increase dementia risk? There are multiple mechanisms.[16]

First, let's take a look at what is happening in the brain during fear and anxiety. Fear activates your amygdala, like a smoke alarm for your brain. Normally, that warning can save your life. But imagine a smoke alarm going off all the time, even when there is no smoke. The blaring alarm would interfere with your sleep and your ability to focus. This unrelenting alarm is essentially what happens when you have an anxiety disorder. When you're afraid, the brain sends electrical signals that lead to the release of chemicals like epinephrine and adrenaline to get your heart pumping: this is the fight-or-flight response. This release of chemicals is normal and protective and is meant to be a *limited* response that ends when the threat ends. If these chemicals are released too often, they can damage cells and DNA, and wear out the brain. In anxiety, there is a chronic release of these chemicals because the fight or flight, or the alarm, stays on.

Next, as mentioned, depression disrupts functionality and communication in multiple parts of the brain. Depression disrupts the lateral orbitofrontal cortex, which is involved in our ability to remember things that made us happy or gave us pleasure.[17] Poorly treated or recurrent depression can also cause excessive stress hormones to be released in the brain, shrinking the hippocampus.[18] This can have a devastating impact on memory, specifically retaining new information.

Last, bipolar disorder is often characterized by a less active prefrontal cortex, which struggles to apply the brakes to control moods, along with a more active and reactive limbic (want) system. In the manic phase of bipolar disorder, individuals may feel like they do not need to eat or sleep and have an inflated sense of self. This puts people at a high risk of exhaustion. A brain cell usually pulsates to a rhythm. Faster pulses are visible in an individual in a manic state.[19] When the cells pulsate too fast for too long, the cell can become burned out, like a battery wears out from overuse.

Mental health disorders, in general, can have the following effects:

- They increase the risk of cardiovascular disease through changes to the stress response system and inflammation. In essence, an overactive fight-or-flight response can wear out the heart and damage brain cells.
- They may isolate us from other people.
- In depressive states, individuals are less likely to engage in physical activity.
- They lead to lack of rest and sleep, which wears on the cardiovascular system.

All these are risk factors for dementia.

The take-home message is that mental health disorders can wear out brain cells, which prematurely ages the brain. An aging brain is at risk for dementia.

Early Childhood Experiences and Brain Aging

In chapter one, we discussed brain development and many of the amazing dynamic processes that take place in the first few decades of life. Could traumatic experiences during that time negatively impact the development, and thus structure, of the brain in a way that raises the risk for dementia later on? One study found that three or more traumatic experiences, such as abuse or neglect in childhood, increased the risk of developing dementia decades later.[20] During brain development, traumatic experiences can disrupt the complete formation of the prefrontal cortex and hippocampus.[21] This disruption can manifest in a brain that has difficulty managing mood. The developing brain is vulnerable and needs to be protected.

Imbalances between chemicals that occur naturally in the brain can also contribute to brain aging. Neurotransmitters such as dopamine and serotonin, which regulate mood and focus, pass between brain cells and are crucial aspects of mental health. For instance, when someone with bipolar disorder is in a manic state, the levels of dopamine and serotonin passing from one brain cell to another are elevated. Mechanisms within the brain notice that the levels of chemicals being squirted out of these brain cells are too high, and shut off their release. (We're not yet completely sure *how* this biochemistry works—just that it does.) The brain goes from being flooded with these chemicals to deprivation. This can lead to a vicious cycle of brain cells releasing too much or too little of these critical chemicals, which can eventually negatively affect memory. Chemical imbalances also factor into depression, where there is evidence that feelings of low motivation are caused in part by lack of dopamine. And, while the role of serotonin is not completely clear in depression, we do know that low levels of serotonin can negatively impact mood.[22]

The role of serotonin and dopamine in mental health has been oversimplified in the past. These key chemicals absolutely play a role in these conditions, but whether their imbalance is the root cause or a secondary effect is not clear. The brain is too complex for us to say that simply adding or subtracting one ingredient will achieve balance. Mental health disorders like depression and anxiety are not caused just by a brain that is out of balance, but that's definitely part of what's going on; in order for memory to function properly, there must be a balance of these key brain chemicals. Too much or too little throws off that balance, damaging and aging brain cells.

Revisiting Inflammation

In the previous chapter, we talked about how inflammation ages the brain. It's worth revisiting the topic here. Remember we discussed a link between moods, the brain, and the body's immune response. Inflammation and autoimmunity also affect mental health. This is something that's particularly important for women to know.

Imagine a woman in her forties to fifties sitting in a doctor's office, detailing her struggle with depression. We can feel her pain and frustration: She has tried antidepressants and talk therapy, yet she hasn't gotten relief from her symptoms. She cannot pinpoint where her depression stems from. She talks about the stress in her life, but it's not enough to warrant her terrible feelings. She has been passed from specialist to specialist on an exhausting, seemingly endless merry-go-round.

In many cases, a doctor might suspect that the cause of her depression is menopause, which seems logical and plausible, as changes in hormones can impact mood. However, in some cases, menopause is not the root cause of depression and/or anxiety.

A study from *JAMA Psychiatry* analyzed 36,000 patients and found that 50 percent of patients who suffered from depression or anxiety also tested positive for antibodies for autoimmune thyroiditis (AIT). AIT, also known as Hashimoto's thyroiditis, is a condition in which one's immune system mistakes the thyroid and thyroid hormones as a threat and produces antibodies to attack and destroy the thyroid, which loses its ability to perform its critical functions. Your thyroid, a butterfly-shaped gland in your neck, secretes hormones that influence a wide range of daily functions, including metabolism, energy production, body temperature, and growth. When it doesn't produce enough hormones, you may experience symptoms like fatigue, sensitivity to cold, dry skin, hair loss, weight gain, and yes, depression and memory issues. It turns out that patient was suffering from an inflamed thyroid. The problem wasn't rooted solely in her brain.

Another interesting twist in the unexpected connection between AIT, anxiety, and depression is that women are impacted by AIT more often than men. And the condition tends to develop between the ages of forty and fifty—just when a physician might incorrectly deduce that postpartum issues or menopause must be the culprit for the unexplained anxiety or depression.

For decades, conventional thinking approached anxiety and depression solely as a disease that involved the brain, so antidepressants and talk therapy were assumed to be the best treatment. When those treatments didn't relieve a patient's symptoms, and the patient was a middle-aged woman, doctors considered menopause as the likely cause and prescribed hormone therapy. (As a side note, a lack of response to antidepressants is not rare: 30 percent of patients with depression don't respond to antidepressant treatment.[23]) They didn't think to check a woman's thyroid or for other autoimmune conditions. As we discussed in chapter eight, whenever a person's immune system attacks the brain, it can destroy critical brain cells involved in memory and raise the risk of dementia—no matter someone's sex. It is important to remember that the relationship between mental health and inflammation is a two-way street in all people. Mood disorders can raise levels of inflammation by elevating stress hormones. And inflammation can raise the risk of depression, anxiety, and dementia by attacking the brain. And a brain that is under attack ages prematurely.

I hope that this chapter impressed upon you how important it is to seek treatment if you're experiencing a mental health issue. It's not just about living a happier, calmer life now (though that is important). Looking after your mental well-being now will also protect your memory and brain function years in the future. Mental health stigma and negative assumptions about mental illness are decreasing, and that's a wonderful thing: With more awareness and public acceptance, more people will get the treatment they need.[24] There is not a one-size-fits-all treatment for these complex mental health conditions, but with careful diagnosis by a physician, there is hope, and these conditions can be managed. Treating depression, anxiety, or bipolar disorder may involve medication and therapy. Living a healthy lifestyle is also a major piece of the puzzle—and those same lifestyle factors that help to support mental well-being are the same critical habits and choices that can protect the brain from the buildup of trash, inflammation, insulin resistance, and memory loss. So with that, let's finally jump into part three: *how* to age-proof your brain.

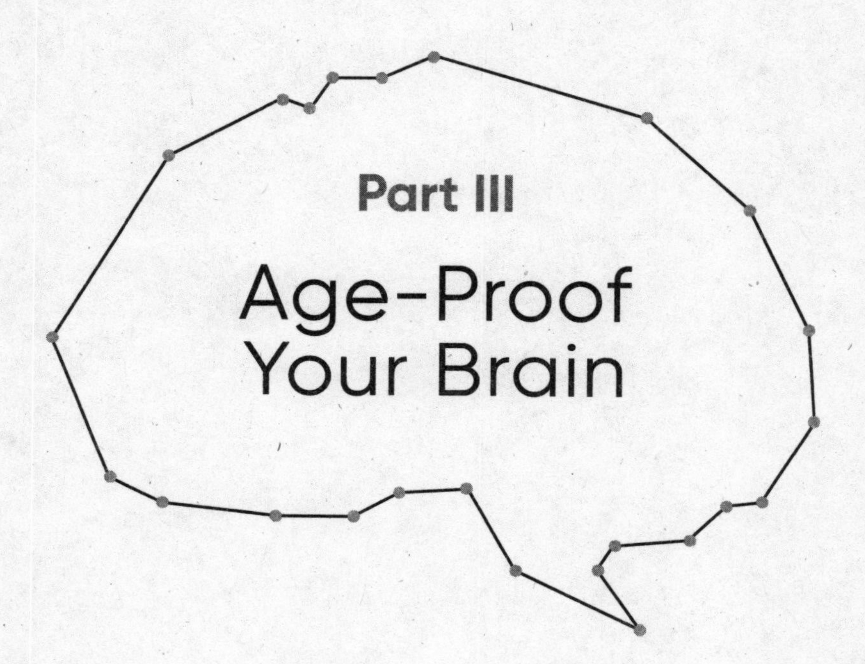

Part III

Age-Proof Your Brain

CHAPTER 10

Sleep: It's Not Just the Number of Hours

PETE SAMPRAS, ONE OF THE BEST TENNIS PLAYERS OF ALL TIME, HELD Grand Slam records, won Wimbledon seven times, and is known for his powerful serve. For much of the 1990s, he was *the* face of tennis. His epic matches with the other tennis greats of his time, like Andre Agassi, Michael Chang,, and Jim Courier, are legendary.

I couldn't tell you what made him such an elite player when it came to the ins and outs of tennis, but I do know something that surely contributed to his success. Wherever Sampras traveled—and he traveled all over the world, crisscrossing multiple time zones—he carried with him a roll of black masking tape.[1]

That tape wasn't for his racket or injuries.

Every time Sampras entered a new hotel room, he took the tape and covered every electronic light in the room. Clocks, fire alarms, cooling units, even the little red light on the TV that stays on when the TV is off. If daylight peeked through the drapes, they were taped closed. Anything that emitted light got smothered by the black tape.

Why? Sampras discovered he played better, was more focused, and had more energy when he slept in true darkness. Back then, when people learned that he was doing this, there was talk that this was over-the-top, diva-type behavior. But now, we know he was on to something. Sampras knew that light—no matter how dim—had the potential to wake him from sleep. He knew instinctively what scientists have since learned through research and studies: artificial light interferes with our natural sleep cycles.[2] And disrupted sleep messes with our brains. In other words, it is not only the number of hours of sleep you get but also *how effective* those hours are.

There's a reason why this chapter on sleep kicks off part three, which is all about brain-boosting and immune-balancing strategies: Sleep is perhaps the greatest ally in the fight to preserve your brain—and quality, deep, restorative sleep is crucial to your brain health.

The Science of Sleep

Sleep is something you might think about when you don't get enough of it or when you wake up asking yourself why your dream just featured an otter, your kindergarten teacher, and a swimming pool filled with ten thousand nectarines. But sleep isn't just rest or time to dream. You might be surprised to hear that sleep difficulty is related to such diseases and conditions as diabetes, heart disease, depression, anxiety, cancer, and obesity, as well as Alzheimer's and dementia.

Poor sleep doesn't put us at risk for these conditions just because it makes us tired. There is more to sleep than just rest. For one thing, lack of sleep also throws off our circadian rhythm, our internal body clock that orchestrates aspects of health such as mood, metabolism, energy, and hormone release. You have probably already noticed how you feel the week after changing your clocks for daylight savings time in the spring. We adjust our clock an hour forward, losing an hour of sleep, and many people lament that they just don't feel right the week after. That feeling manifests in the world: there are more traffic accidents and workplace injuries, as well as more heart attacks and strokes, the week after daylight savings time.[3] A study at the Columbia business school even found judges give out harsher rulings the week after changing to daylight savings.[4] (Now, that's a pro tip if you ever need to schedule a court appearance!)

So why is sleep so crucial for our bodies and brains? Let's take a closer look at what happens when we nod off at night.

Understanding the Sleep Cycle

We cycle through three stages of sleep: light sleep, deep sleep, and rapid eye movement (REM) sleep. There are some complexities to this cycle, but we can essentially boil it down that way. Those three stages take about ninety minutes total, and you keep repeating this cycle all night long.

One way to recognize each stage is by measuring the amount of electrical activity in the brain. Your brain cells communicate by sending electrical signals to each other. The harder your brain is working, the more electricity your brain generates:

- In light sleep, the amount is similar to the amount when you are awake.
- In deep sleep, there is very low electrical activity relative to light and REM sleep.
- In REM sleep, there is a higher level of electrical activity than when you are awake.

In a normal sleep cycle, your brain wakes up every ninety minutes, even if you're not aware of it. That's why you shouldn't worry if you wake up briefly during the night—it's a natural part of the sleep process. In fact, knowing that waking up is normal is the first step in getting effective sleep. Where we run into trouble is when we wake up and start thinking about all the things we didn't do yesterday and what we have to do tomorrow, or we are so stressed about waking up that we can't get back to sleep. If you do wake up fully, avoid looking at the clock or thinking about tomorrow's to-do list. Take several deep breaths and remind yourself your brain simply finished a cycle of sleep and now it's time for the next cycle.

Chill Out for Better Sleep

Have you ever tossed and turned trying to sleep in a room that was too warm? Most people find it easier to fall asleep in a cool, rather than warm, room, and there's a reason for that. To see why, let's explore your sleep cycle a bit more. The first phase of the sleep cycle, light sleep, lasts about twenty to thirty minutes. During light sleep your brain has significant electrical activity, which is why if you wake up in this phase of the sleep cycle, you will feel most refreshed. Your brain lowers your core body temperature to transition from light sleep to deep sleep. A study found that if you are having trouble falling asleep, lowering the temperature of your bedroom a couple of degrees helps you reach deeper, brain-boosting sleep.[5] To reduce the temperature of your room, lower the thermostat or open a window slightly. You can also look into lighter or more breathable sheets, blankets, and pajamas.

We want to make sure we are doing whatever we need to do to get several cycles of deep and REM sleep throughout the night. Deep sleep is the anti-aging, regenerating, and rejuvenating body repair phase of sleep. Brain activity slows down significantly, as this is the time your brain, as we discussed in chapter two, wrings the trash out of 80 billion brain cells, before fluid from your spinal cord floods your brain to wash it away. While your brain is being deep cleaned, your body is also repairing itself by rebuilding muscle and bone. If we don't get effective deep sleep, we age significantly faster. Plus, as we get older, we can get less deep sleep, which can wreak havoc on the brain and immune system. Deep sleep begins to significantly decrease starting around the age of thirty.[6] This lack of effective sleep as we age is a key factor in memory and brain and immune issues.

Sleep and Your Immune System

Sleep is a secret weapon for your immune system. Researchers wanted to know if people were more likely to get a cold if they didn't get a good night's sleep, so they took a group of volunteers and sprayed cold virus in their faces. (I repeat: these were volunteers.) Half the group got to experience uninterrupted deep sleep, and the other half were awakened each time their brain entered deep sleep. Those whose periods of deep sleep were interrupted caught significantly more colds.[7]

A similar study found sleeping four hours a night carried the greatest risk of catching the common cold. Insufficient sleep was a greater factor than age, stress levels, or being a smoker or nonsmoker.[8] During sleep your T cells fight off viruses, bacteria, and even cancer cells. Sleep allows the T cells to activate sticky proteins called integrins.[9] These integrins help T cells attach to these invaders or infected cells and kill them. If the T cells cannot attach and destroy the dangerous pathogens, we are far more likely to become sick.

The REM, or dreaming, phase of the sleep cycle is critical to remembering all the things you learned during the day. As the connections between your brain cells are currently reminding you, everything you learned is housed in connections between your 80 billion brain cells. When you dream, your brain runs electrical stimulation over those newly made connections. Whether you are studying for a test or just want to remember what you discovered in that last golf lesson or French class, if you don't have REM sleep, you don't learn as effectively.

In addition to strengthening new connections during REM sleep, you are unhooking any old connections your brain thinks are no longer important. That is why your dreams are often a bizarre mix of things you just learned or experienced and people and things you haven't thought about in a long time.

REM Sleep Disorder

While we are dreaming, an area called the pons at the back of our brain sends a signal down the spinal cord to paralyze the body from the neck down. This is normal. Why would this happen? If you were not paralyzed, you would act out your dreams—which could be very dangerous if you were dreaming about something like a fight. If a person is not paralyzed during dreaming, it is called REM sleep disorder. REM sleep disorder is *not* talking in one's sleep or some slight movement; it's also different from sleepwalking or the side effects of sleep medications that cause some individuals to get out of bed and eat or walk around while still asleep. REM sleep disorder is actually getting out of bed and *acting out* a dream. There is evidence that this disorder can be an early warning sign of some brain diseases such as Parkinson's and Lewy body dementia.[10] This does not mean that if someone has REM sleep disorder, they will definitely develop those conditions. It just means they have an elevated risk, and that this is a warning sign that should not be ignored. If you are experiencing any of these sleep abnormalities, it is important to make an appointment with a neurologist.

Now, to better understand how you can use sleep to keep your brain and immune system young, let's take a deeper dive into breakthroughs that can help improve sleep.

Your Brain Clock

A small clump of twenty thousand brain cells called the suprachiasmatic nucleus, or your "brain clock," is so important to our understanding of the brain that the people who discovered how the brain clock works (Jeffrey C. Hall, Michael Rosbash, and Michael W. Young) won the Nobel Prize in Physiology or Medicine in 2017. This cluster of cells, the size of the head of a pin, is one of the most exciting areas of medical research because it's the master clock for your entire body.

Your body is made up of about 37 trillion cells, and many of these 37 trillion cells are listening to that tiny clump of 20,000, which sets the rhythm and pace for what all the other cells need to do: heart cells to keep your heart beating, stomach cells to digest your food, adrenal cells to release hormones into your blood. For a simple picture of how your brain clock works, take the following scenario. You're in bed at night and all the lights are off. Here's what happens in your brain:

1. Your brain clock notices the darkness and causes your brain to release a chemical called melatonin.
2. Melatonin puts your brain to sleep.
3. In darkness, melatonin continues to be released, keeping your brain asleep.

When it's time to wake up:

1. Natural light from outside enters through windows and passes through your eyes. This occurs even if your eyelids are closed.
2. The light tells your brain clock to turn off the melatonin.
3. In the absence of melatonin, you wake up.

But just as a bad conductor can throw off the rhythm of an orchestra, a malfunctioning brain clock can derail your heartbeat, breath, metabolism, mood, digestion, and sleep. For example, older individuals without memory issues can check into a hospital and check out with memory issues.[11] Why? In hospitals, the lights are often on all night long, and patients are frequently disturbed by nurses taking vital signs, or by noise from a roommate or in the hall. These seemingly insignificant disruptions at night, coupled with lack of natural light during the day, wreaks havoc on effective sleep and memory preservation.

In today's world, we have a lot of disruptions in both the "fall asleep" and "wake up" phases of sleep. Your brain evolved over millions of years, but it hasn't changed much in the last ten thousand. It hasn't caught up and adapted to the modern twenty-four-hour lifestyle. The good news is that we can use our brain clocks to overcome the effects of sleep disruptions. Very simple

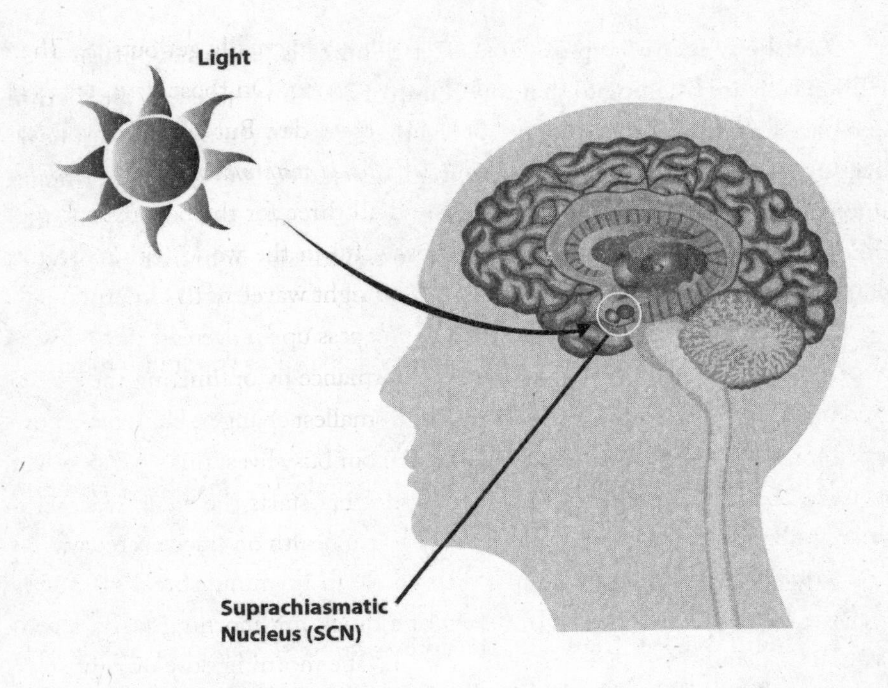

adjustments to return your brain to its natural rhythms of natural light and true darkness can make a big difference for the better.

Use Light to Set Your Inner Sleep Timer

Your brain clock is like a timer you can set to help you fall asleep. How? After waking up, get outside in natural light for about ten to fifteen minutes. Yes, we start preparing for sleep when we wake up in the morning. A quick walk around the block will do the trick. Once you do that, you start the countdown to falling asleep at night. It sounds simple, but missing those few minutes of natural light can upset your brain clock. If you work from home, or jump in a car and go straight to an office, you may miss this dose of natural light and set yourself up for sleep problems.

I'm often asked, "Do I have to go outside? Can I stand by my window? Can I put a treadmill by a window? Is there anything I can do to not go outside?"

Look, I understand that it's not always easy to actually get outside. There are days when it is too cold or difficult or it's raining. On those days, try to be near windows in the morning and throughout the day. But whenever you can, get outside. *A key point is that natural light, coupled with movement and fresh air, is a trifecta of goodness to the brain.* You need all three for the best results. And you don't need the day to be the most beautiful in the world, either. Studies also show that even the cloudiest day has the right wavelength of natural light to stimulate the brain clock correctly, so don't pass up an overcast day.[12] I work with professional athletes to maximize performance by optimizing their sleep, and this tip to get morning light is one of the smallest changes, but it has a powerful impact. Though it is often overlooked in our busy lives, this tip is powerful because the morning light is like the domino that starts the chain reaction to keep multiple aspects of our mental and physical health on track each day.

It is also important to take two to three 10- to 15-minute breaks throughout the day and get outside. Think about a quick midmorning walk, a walk after lunch, and then a later afternoon stroll. The morning dose of light starts the countdown, and the doses of natural light throughout the day keep your brain clock running smoothly and on time.

Another common question is, I wake up before the sun comes up, so what should I do? There is nothing wrong with waking up early if you're getting enough sleep, but once the sun comes up, take a quick walk outside. For all the late risers, waking up later can work—it's just important to be aware that waking up *too* late can push your bedtime later. There is evidence that going to bed past midnight or 1 AM can raise the risk of mental health issues such as depression, but going to sleep just an hour earlier can lower the risk of major depression by 23 percent.[13]

An interesting study looked at a group of people who had great difficulty sleeping.[14] These individuals attempted a variety of techniques, including sleeping pills, to reset their brain clocks, but everything they tried failed. The researchers took them camping, which meant they got plenty of natural light and experienced true darkness—no electronics!—at night. Within a weekend, the participants got their brain clocks back on schedule and slept significantly better.

You can try this. (If you do, don't go glamping, and don't bring your tech and every modern convenience with you.) If you don't want to sleep in a sleeping bag or a tent, just take an honest look at your day. Ask yourself:

- How much natural light am I really getting in the morning and throughout the day?
- How much true darkness am I getting at night?
- Am I in the presence of a lot of artificial light in the evening?

Start making small steps to adjust wherever you can. Consider lowering the lights and limiting the number of hours in the presence of artificial light before bed. The following pages share my best tips for brain-boosting sleep.

"Are You Still Watching?"

Beware of binge-watching TV at night. There is a rumor that at a Netflix meeting, the company's CEO announced to employees that their biggest competitor was sleep. Streaming services utilize brain-science-based tricks to keep us glued to the screen. One trick is starting the next episode before the end credits roll. Before you know it, you are hooked into the next episode, and then "one more episode" turns into the entire season before sunrise. Remember, a regular sleep schedule and quality sleep are critical ingredients in balancing the immune system.

How to Get the Best Sleep for Your Brain

I don't think anyone would argue that we're not in a sleep crisis. Besides the fact that our work lives put pressure on us to grind around the clock, we're also being distracted by technology that can encourage us to put off sleep.

In addition, pervasive health problems that have symptoms such as pain and anxiety simply make sleeping harder.

Unfortunately, we've relied too heavily on sleep medications. A study in *JAMA Internal Medicine* uncovered that over-the-counter cold, pain, and sleep aids, or medications that induce drowsiness when taken regularly, can significantly raise the risk of dementia.[15] These sleep aids put the brain in a trancelike state that is different from effective deep sleep, and do not allow the brain to effectively wash and remove brain trash. That's why sleep medications are no longer recommended by the American Academy of Sleep Medicine as a first line of treatment for insomnia.[16]

Even melatonin supplements are probably not a good idea. First of all, they're not regulated by the FDA. A study published in the *Journal of Clinical Sleep Medicine* tested the amount of melatonin in different brands of the supplement. It turns out there was either too much, too little, or completely different compounds in each of the supplements compared to what was advertised on the label.[17] (This is a common problem in any supplement, melatonin or not, that's not regulated by the FDA: the reality is that you don't always know what you are getting.) Melatonin supplements can also interfere with other medications and even stop the brain from making its own melatonin, or what is called endogenous melatonin release. The caveat to these recommendations is whether your personal physician has determined that melatonin supplements are right for you.

The long and short of it is that rather than relying on external aids (unless your doctor specifically tells you otherwise), use what we know about how your brain works to get effective sleep.

To Fall Asleep . . .

Take a warm shower or bath. An analysis of 5,322 studies found there is something very simple you can do that can help you fall asleep faster, in addition to getting a better night's sleep: Take a warm shower or bath ninety minutes to two hours before you get into bed.[18]

Yes, a warm shower or bath is relaxing, but there is more to it. In order for the brain to fall asleep, the core body temperature needs to drop, as we discussed in the section on sleep cycles. If you take a warm shower or bath about ninety minutes before bed, your body will cool itself after being warmed up from the water. This process will lower your overall body temperature and you'll feel cool enough to fall asleep. The time frame is ninety minutes because this gives the body enough time to adjust and be at the right temperature for sleep. A warm shower helped people fall asleep ten minutes faster, as well as get a more restful and effective night's sleep.[19]

Keep paper and pencil next to your bed. Often a stressful thought will keep us from falling asleep or will wake us up in the middle of the night.[20] A surprisingly simple solution is to write down a list of things you are worried about and tasks you didn't get to.[21] (Remember to use old-fashioned pen or paper to avoid looking at a screen before bed.)

In one study, when people did this simple practice before bed, they fell asleep faster and slept significantly better than those who did not write down their to-do lists.[22] Why? Think about your phone contact list and how many phone numbers you have memorized now, vs. fifteen or even five years ago. You don't have to remember phone numbers because your phone does it for you. When your brain knows important information is stored safely, it lets that information go. Interestingly, this study found that those who made their to-do list more detailed fell asleep even faster than those who made a simple list. Use this trick to your advantage and store your stresses, worries, and to-do list on paper, tell your brain it's stored safely and you can revisit it in the morning, and then get a good night's sleep!

Power down smartly. Imagine you were driving eighty miles per hour on the freeway and pulled into your garage at that speed when you arrived home. That wouldn't be wise. Yet that's what we do when we multitask right before bed, checking our phones, finishing up work, attending to last-minute chores. Create a buffer between your day and sleep. Don't fall into the trap of thinking that if you just get one more thing done, you will have a less stressful day tomorrow. Because without a good night's sleep, you *will* have a stressful day tomorrow.

Sleep is the time your brain recalibrates its stress response and recharges the parts of your brain that calm you down. Research presented at the American Academy of Sleep Medicine has shown that fifteen minutes of mindfulness (we'll talk more about that in the next chapter) can help sleep quality.[23] Relaxing activities like light stretching, listening to music, journaling, making that to-do list or list of worries, or reading a book have also been shown to be effective.[24] Just make sure you do a relaxing activity so you slow down between your hectic day and a peaceful bedtime.

Establish a routine. Have you ever driven from work to home or home to work, gotten to your destination, and asked yourself, "How did I get here?" Even when you're distracted, your brain notices the familiar visual routine and goes on autopilot. For example, your brain unconsciously notices that a specific street sign follows another street sign that you see all the time. The last thing your brain wants to do is waste energy, so when it recognizes this pattern, it figures it doesn't have to exert itself with conscious thought.

Use this brain trick to your advantage and pick something that you will do every night about a half hour before bed. A few options are light stretching or writing down your worries, as we mentioned in the previous section. I have had people tell me they have been trying to read the same boring book for ten years and that puts them to sleep. That's a perfect cue to the brain that the day is over and it's time to rest. But if you don't want to read the same page over and over, just think of activities that can be done in low light and that are not overly stimulating. These relaxing activities will not only calm you but will also cue your brain that sleep is next, and your brain will go into autopilot to put you to sleep.

Create a sunset in your bedroom. For many people, their pre-bedtime routine looks like this: all the lights in the bedroom and bathroom are on, and the TV is turned to the news while they're brushing their teeth. They turn off the lights and TV and jump in bed and wonder why it's taking them a half hour to fall asleep.

It is because that clock in the brain is tuned to not fall asleep right away. For most people, the time between their brain clock triggering the release of melatonin and actually falling asleep lasts twenty to thirty-five minutes. If you want to increase your chances of falling asleep faster once you are in bed, create a sunset in your bedroom. This lets your brain clock know that it's nighttime so it will begin the process of releasing melatonin. Simply turn off the TV and lower the lights a half hour before you want to fall asleep. And if you're going to read, don't do it on an electronic device. A study published in the *Proceedings of the National Academy of Sciences* had two groups of people read the same story before bed, but one group used a light-emitting device like a smartphone or tablet, while the other group read a physical book. The second group fell asleep faster and slept better.[25]

Those electronic devices we love release blue light, which is similar to sunlight. When you use your smartphone, tablet, or computer, your brain thinks it's daytime and delays the release of melatonin. (Yet another reason to put down your phone.) You might be wondering about those apps that claim they can block the blue light. Unfortunately, none of them have been studied enough by anyone other than the companies that sell them. Also keep in mind that the information on the device itself can be stressful or linked to stress. For example, if you're looking at your work phone in the hour before bed, your brain may associate your phone with work stress. That alone can disturb sleep. A study published by the American Academy of Sleep Medicine found that people who looked at their phones in the hour before bed tended to be hungrier and ate extra calories before bedtime.[26] You wouldn't think these two activities were related until you thought about the relationship between blue light and daylight and eating. Tricking the brain into thinking it is daytime can also turn on mechanisms related to hunger, as we are programmed to eat during the daylight hours.

Hydrate at the right time. Overall, the National Sleep Foundation recommends that, based on body weight, most women need approximately 91 ounces of fluid daily and most men need about 125 ounces. Dehydration can have a negative impact on sleep. Parched nasal passages can increase the loudness of snoring, waking you and anyone within earshot. If your muscles don't have enough water, they may cramp, and the pain may wake you.

The solution isn't to chug water before you go to bed; waking to go to the bathroom several times a night isn't good for your sleep, either. It's all about spreading out the amount of water you need throughout the day. If you wake up frequently to go to the bathroom, stop consuming fluids about two hours before bed. In certain circumstances, where you might *need* to drink water before bed, consider hydrating throughout the day so that the amount of water you drink right before bed is reduced to a small glass.

Sleep in true darkness like Pete Sampras. You probably think you sleep in the dark, but do you? Tonight, take a look around your bedroom. Is it really dark, or just modern dark? Our bedrooms are now filled with charging cell phones, night-lights, or light from a television, streaming device, or computer. We now know these little bits of light add up to light pollution that can keep your brain from reaching the deepest, brain-boosting levels of sleep. Unplug these devices before you get into bed, or move them to another room. You could even invest in blackout shades and curtains. If you want to pull a Pete Sampras, use black tape to cover any little lights you can't remove.

A 2022 study investigated if even small amounts of light present in the bedroom had an impact on health.[27] The amount of light used in the study was not enough to read with. The participants reported that they didn't notice the little bit of light while they were sleeping and felt it didn't disrupt their sleep. Yet tests done of the subjects revealed a different story: elevated heart rates and insulin resistance were found throughout the night and the next day in those that slept with a small light on. The nervous system was activated throughout the night in those who had this little bit of light, and this had a negative impact on cardiovascular and metabolic health. The take-home message is, even if you don't notice the light, your body does.

There are some lights, such as streetlights or neighbors' lights, that we cannot control. In those cases, experiment with a light-blocking window covering or a sleep mask or clean towel that covers your eyes. Interestingly, research has uncovered that certain cancer medications work better if the patient sleeps in true darkness, highlighting the power of an optimized brain clock in the healing process.[28]

When You Wake Up and Throughout the Day . . .

Set your alarm strategically. When you wake up, the phase of the sleep cycle you are in will dramatically affect how you feel. If you are woken from deep or REM sleep, it's like waking up on the wrong side of the bed. Ideally you want to wake up when you're in light sleep; that's when you will feel the most energized and refreshed. One way to do this is to figure out the number of hours of sleep you need and set your alarm clock about fifteen to twenty minutes after when you calculate you will complete your cycles naturally. Fifteen to twenty minutes will give you a nice buffer of time to try to wake up on your own.

While this won't work perfectly every day, the more days you can wake up in light sleep after completing your cycles, the better you feel.

Get out early. Remember: you prepare for the next night's sleep by getting outside in natural light first thing in the morning. Spend ten to fifteen minutes outside every morning, and you will be surprised how setting your brain clock helps you fall asleep at night. Walk your dog, check the mail, or just take a stroll around the block.

Keep your brain clock on time. Throughout the day, try to be near windows to keep your brain clock aware that it is daytime. Try to take a midday walk outside around lunchtime for another dose of light for your brain clock to keep running smoothly.

Nap smart and keep it under thirty. I'm all for the power nap, but you have to do it right. You'll get the most out of your nap if you sleep no more than thirty minutes, the amount of time of your light sleep cycle. If you sleep longer, you enter deep sleep and you will awaken groggy and disoriented. Take a tip from professional athletes and set a timer for thirty minutes to get a brain-boosting, energy-maximizing nap without going into deep sleep. NBA teams such as the Chicago Bulls schedule time for the entire team to take a quick nap before a game to optimize energy, recovery, and focus.[29]

Still Feeling Groggy? Get Checked for Sleep Apnea

There is a critical link between sleep and memory. People with untreated sleep apnea raise their risk of memory loss on average ten years before the general population.[30] If someone suffers from sleep apnea, they wake up hundreds of times during the night, which disrupts the sleep cycle and the memory-strengthening process of reinforcing brain cell connections during sleep. An individual with sleep apnea can be waking up every two minutes and not even be consciously aware of these disruptions, even as they take a huge toll on the person's memory. If sleep apnea is effectively treated, those memory issues disappear.

Symptoms of sleep apnea can include loud snoring, but snoring does not necessarily mean someone has sleep apnea; conversely, you may have it and not snore very much at all. One often overlooked symptom of sleep apnea is the feeling that sleep was not restful or refreshing. If the tips discussed in this chapter do not promote restful sleep, it is important to determine if sleep apnea is a factor. Sleep apnea can be caused by a variety of reasons, from weakened musculature in the neck to misregulation in the brain. The specific treatment is based on the underlying root cause of the apnea; thus it is important to determine the root cause by discussing your sleep issues with your personal physician and possibly visiting an accredited sleep center for further evaluation.

Your FAQs About Sleep

These are the most common questions I receive when I give presentations on sleep.

How Much Sleep Do I Need?

It seems that every time you read an article about the number of hours of sleep you need, you get a different figure. Here's the bottom line: for most people, a healthy brain needs somewhere between seven and nine hours a night, but the number of hours of sleep *you* need is specific to *you*. There is a rare group of people who can function at high levels both physically and mentally on significantly less sleep, such as four hours a night. These are called short sleepers, and it is believed there are genetics involved.[31] *True* short sleepers likely account for less than 1 percent of the population.[32] Some people who think they are short sleepers are really just sleep deprived—and thus vulnerable to all the issues that come with lack of sleep. If you believe you might be a short sleeper, it is important to have an honest assessment to determine whether that's really the case.

You need the number of hours of sleep that you get when you wake up in the morning feeling refreshed and energized for the day. There *is* such a thing as too much sleep: If you're sleeping more than nine hours a night, you should be evaluated by a personal physician. Too much sleep can be a sign of conditions such as heart disease, diabetes, and depression, and can raise the risk of memory issues, back and neck pain, and obesity.

So how do you figure out how many hours of sleep you need? Find a time when you can do the following things for three to four days:

- No caffeine and alcohol. Caffeine has a half-life of about five to six hours. That means that after five to six hours, only about half of the caffeine you've ingested will have left your system. Thus, if you're having trouble sleeping, stop all caffeine intake—coffee, tea, soda, energy drinks—by 2 PM. If you aren't having trouble sleeping, there is evidence a moderate amount of caffeine in coffee and tea might have protective elements for brain health.
- No electronic devices for two hours before going to bed.
- No sleeping aids, either prescription or over the counter.
- Go to sleep when you get tired.
- Don't set an alarm clock.
- Get up when you feel rested.

Keep track of the number of hours you slept over these four days; that's the optimum amount of sleep your brain and body need.

Now it's time to take steps to get that optimum amount of sleep in your regular routine. (Unless you've discovered you're naturally an early riser, your alarm clock probably needs to come back into the picture.) You may not be able to wake up when you want to on a workday, but you can adjust your bedtime and implement a consistent routine that will help you get to sleep. As part of your new routine, do the following:

- Turn off your devices an hour before bedtime.
- Create a sunset in your bedroom.
- Before bed, do something relaxing, such as listening to soft music, light stretching, or mindful breathing exercises.
- Write down your anxious and stressful thoughts on paper.

What if I Can't Fall Asleep?

Let's say you've gone through your bedtime routine, gotten under the covers, and twenty to thirty minutes have gone by but you're still not asleep. Should you tough it out or get out of bed? Remember that your brain is a pattern and association machine, and the last pattern and association you want to set up is that your bed is a place where you don't sleep, or a place where you toss and turn and worry. This is why we want to be very careful about what activities we do in bed. Avoid stressful activities, like doing your taxes, in bed.

If you can't sleep, get out of bed. Keep the lights low and stay away from your devices. Listen to some relaxing music or do some light and easy stretching like trying to touch your toes. If something is bothering you, write it down. This is the time to read a boring book or article. After about fifteen to twenty minutes, go back to bed and try to fall asleep.

What About Having Alcohol Before Bed?

A nightcap or before-bed alcoholic drink is often thought of as a sleep aid because alcohol is very good at putting people to sleep. The brain makes a stimulant called glutamate, which keeps you awake, but alcohol destroys glutamate, making you feel drowsy. The problem is, about four hours after alcohol destroys the glutamate, your brain wonders, *Where did all that glutamate go?* So it makes more glutamate. In other words, if you had a drink at 10 PM, you'd start making a stimulant at 2 AM, which would wake you up and ruin your sleep. For this reason, alcohol should not be used as a sleep aid. It isn't one.

Should I Listen to My Sleep Tracker?

A lot of people ask me what to do if their digital sleep monitors or apps tell them they're not getting deep or REM sleep. Before we go any further, we have to answer these questions: Are sleep trackers accurate? Are they giving you any information regarding sleep that could be used to make a diagnosis? No and no. This is an important point, as people can get so stressed about information their sleep tracker is giving them—but it's not accurate information to begin with. Sleep trackers are an example of where marketing is ahead of science. The only way to accurately determine which phase of the sleep cycle you are in is to go to an accredited sleep center and have an overnight sleep test.

How Do I Get Over Jet Lag?

Jet lag happens when your brain clock thinks it's one time, but your body is in a different time zone. A new time zone is a sudden shift that throws off the timing and tempo of 37 trillion cells in your body—that's why jet lag feels so awful. But, as you already know, melatonin supplements—which you may have heard were good for jet lag—are no longer recommended for the general population. So, instead of relying on pills or tablets, when you arrive in your new city or return back home, right after you wake up, get outside in the presence of natural sunlight for ten to fifteen minutes. It's the safest, most effective way to reset your clock and get your orchestra of cells back playing on the correct rhythm. Another quick trick if you are jumping several time zones: several days before your trip, start moving your clock about fifteen minutes each night to the time zone you are traveling to.

S.L.E.E.P. Cheat Sheet

These secrets lead to memory-boosting, immune-enhancing, anti-aging, brain-washing sleep, and an end to those frustrating, sleepless nights. It's all about SLEEP:

S: Schedule. That little clock in your brain wants to run like a Swiss clock. A consistent bedtime and wake up schedule is key.

L: Light. Create a sunset in your bedroom before bed and avoid technology. Get out in natural light as soon as you can after waking up.

E: Exercise. Exercise is one of the best things you can do during the day to sleep better at night. We'll cover exercise in chapter twelve.

E: Eating. We are what we eat, and we sleep based on what we eat and drink. Chapter fourteen breaks down food and beverages.

P: Patterns and Practice. You will establish new, healthy patterns of sleep if you practice the tips in this chapter.

CHAPTER 11

The Stress Surprise

WE HEAR IT ALL THE TIME: STRESS IS BAD. IT'S TRUE THAT TOO much stress *isn't* a good thing, physically or psychologically. So, should we strive to have a stress-free life? While this may be a surprise to some people, the answer is no. *Some* stress is good for you. The right kind of stress motivates and focuses the brain and can even slow down brain aging.

Beneficial stress is manageable and occurs in a burst, or a moment. Manageable stress cleans up brain trash; it turns on a program called autophagy, which cleans up old, dead cells and removes waste and toxins, keeping your body clean and young. Think of a challenge you want to tackle or something that makes you nervous—that's the type of stress that grows new brain cells in the hippocampus. Just like the car that we compared your brain to earlier, if you don't "drive"—if there's no stress at all—it falls apart. On the other hand, if you keep asking it to overperform—if you're constantly on edge—it will break down. As I described in chapter four, stress releases the hormone cortisol, which in small, sporadic doses is very good for you, but with prolonged exposure can lead to chronic inflammation (and raise our risk of memory loss, mood changes, depression and anxiety, and dementia).

Here's the big question: How do we find just the right amount of stress to keep our brains and immune systems running like finely tuned Ferraris? Well, that comes down to individual differences and what happens in your brain when you get stressed.

Your Brain on Stress

In chapter five, we talked about your prefrontal cortex (PFC), the part of your brain that calms down your stress response and suppresses stress, anger, and anxiety. To dig a bit deeper into that, the PFC is like a muscle that must be trained and worked out each day to keep it in shape. I wish I could go to the gym on January 1, work out that one day, and I would be good for the year, but unfortunately, it doesn't work that way—and neither does working out the PFC. To keep the PFC strong, we must work it out regularly. What works to strengthen your PFC and manage stress? Have you ever had anybody tell you, "Just calm down"? How well does that work? Not so well, right? Have you ever had someone yell at you, "Just relax"? That's not a winner, either.

What *works*—what *does* manage stress—is to boost happiness.

Boosting happiness minimizes and lowers anxiety, depression, and pain symptoms.[1] Even small increases in happiness extend longevity.[2] It's not easy to completely define what happiness is, but in these studies, it is generally defined as being in a good mood, having positive emotions and enjoyment, and being satisfied with life. Research found the risk of dying due to any cause lowered by 19 percent for happy older people.[3] Just like it doesn't often work to say, "Just relax," it also isn't always effective to have someone tell you, "Just be happy," even if it's printed on a t-shirt or a coffee mug. Both relaxation and happiness are important to managing stress and both can be attained through effort.

How do we find the things that scientifically boost happiness and sustain it? Can you teach people to be happy? How do you have a happiness intervention? Does asking question after question boost happiness? OK, that last one was a joke, but happiness can be tricky, because we often think, *I'll be happier if I just accomplish that goal or obtain this item*. We get that item, or we reach that goal, and the happiness is fleeting, or we don't feel as happy as we hoped. Boosting happiness is about more than that.

A study of 18,000 individuals from all walks of life—young, middle-aged, older, every social class, different countries—tested how to get and sustain happiness.[4] Participants received a text message sent randomly during the day. It asked participants to respond to three questions:

1. How are you feeling right now on a scale of 1 to 10 (1 being terrible, 10 being fantastic)?
2. What were you doing right before you got this message?
3. Were you focused on what you were doing, or was your mind wandering?

They found that 50 percent of the time, people's minds had been wandering. What does that mean? Half of you aren't really paying attention to the words on this page. But for the other half of you who are still with me, the idea is that there was a surprising link between mind-wandering and unhappiness. You would think if we're not happy with what we're doing, our mind would wander to happy thoughts and tropical vacations. Instead, in an overwhelming majority of the cases, people's minds wandered to unhappy, stressful, and anxious thoughts. Much of our stress comes from worrying about the past or the future; the key is that being present in the moment is associated with greater happiness and less stress. So how do we train our brains to be in the present moment? It turns out we can practice being happy. A quick brain science lesson courtesy of Leonardo DiCaprio will show us how this works.

Life Lessons from Leo DiCaprio and the World's Happiest Man

Leonardo DiCaprio played business magnate and pilot Howard Hughes in *The Aviator*. Hughes suffered from obsessive-compulsive disorder (OCD), which has a particular type of brain activity. DiCaprio showed up to work every day and pretended to have OCD. He spent his time doing what people who have OCD do, and his brain changed as a result. He had to work with a UCLA psychiatrist to retrain his brain and get over his OCD impulses.

DiCaprio's experience shows us that any time you learn something, some subset of your 80 billion brain cells reaches out and makes a connection. That's the learning. When you practice that thing that you learned—it doesn't matter if it's a new song on the piano, playing tennis, or learning some brain science—your brain runs electrical stimulation over those same connections. When you repeat what you've learned, you strengthen those connections. Like we discussed throughout this book, the more you do something, the more you strengthen the connections. If you don't practice, those connections weaken, and you become less skillful, or you might forget the stages of the sleep cycles we discussed in a previous chapter. If you practice again, those connections strengthen.

Our emotions function the same way. If we spend our time in certain mood states, we strengthen the brain cell connections associated with that mood. If we practice being unhappy throughout the day, we become an expert at being unhappy. We can even get good at being chronically stressed. On the other hand, if we practice being in positive mood states, we strengthen the connections for positive mood states, and we get better at being positive. Since we are spending less time being unhappy, we get rusty at being unhappy.

When people are happy, their brains look a certain way. One study tested thousands of people and repeatedly found the same pattern: happy brains have more blood flowing to the prefrontal cortex. One subject had a tremendous amount of blood flowing to his prefrontal cortex. Was this the world's happiest man? Well, he said he was very happy. Although it is impossible to objectively measure happiness, as it turned out, he was a Buddhist monk.[5] What do Buddhist monks do all day? They meditate.

What Is Meditation?

Meditation trains the brain to be focused, to be present. There are several types of meditation, but an essential aspect of the practice is *mindfulness.* Mindfulness places the brain in the present moment and asks that whatever feeling the person has in this moment is looked at from a place of acceptance. Isn't it interesting that the idea of *being in the present moment* came up in the happiness study with the text message survey and again in this brain scan study? More evidence that being in the present is critical for happiness and stress management.

After seeing this study, scientists worldwide started taking Buddhist monks and sliding them into brain scanning machines. They kept finding the same pattern. (I would bet there is a monk in a brain scanner somewhere right now.) At this point, you might be thinking what I thought when I first read these studies: *That's interesting, but let's be realistic. Monks don't have my problems or my bills. Do they even pay taxes?* Put that aside for a moment, though, because I want to show you how understanding what's going on in a monk's brain gives us insight into how we can deal with our stresses.

The scientists asked how many hours of meditation the monks did to get to that happy brain state—and found that the answer was thirty-four thousand hours. That's eleven and a half years of meditating eight hours a day. If you have anywhere you need to be in the next eleven and a half years, this is not good news. You would have to really clear your schedule.

Is there a CliffsNotes version? Do you have to meditate for thirty-four thousand hours to get a happy brain? The good news is no, you don't. You can reap the same benefits through practicing mindfulness.

Mindfulness

Mindfulness is the part of meditation that involves being in the present moment, but meditation is just one way to practice mindfulness. In a moment, we will discuss other ways to practice it, but first let's look at a Harvard study that took people who had never before practiced mindfulness and had them do it for eight weeks. The study took pictures of participants' brains before they started, during the eight weeks, and after the eight weeks.[6] After practicing about thirty minutes of mindfulness a day for eight weeks, the participants' brains looked more like a monk's brain. The hippocampus—the part of your brain that allows you to learn new things—grew. The PFC got stronger and larger, and the amygdala, which manages our fight-or-flight stress response, shrank. People who practiced mindfulness still had stress, but they were better able to control and manage it. This highlights that we have the power to change our brains, mood, and how we respond to stress.

Another study, published in *Psychoneuroendocrinology*, showed that people practicing about twenty-five minutes of mindfulness a day had a detectable drop in their cortisol levels after just three days.[7] If twenty-five minutes seems like too much, studies published in *Consciousness and Cognition* found just ten minutes of mindfulness a day also had benefits, relieving stress and improving focus.[8] While chronic stress can cause inflammation, a study published in *Frontiers in Immunology* uncovered that meditation, yoga, and mindfulness lessen inflammation by increasing anti-inflammatory cells and lowering inflammatory immune cells.[9]

There are many mindfulness exercises, some of which involve breathing, some of which don't. If you like breathing exercises, here's a simple and effective one that can be done any time you have a spare minute. (We cover mindfulness exercises that do not involve breathing below.)

Say to yourself, or out loud: "Breathe in calmness, breathe out anxiety."

On the "breathe in calmness," inhale through your nose.

On "breathe out anxiety," exhale through your mouth.

Try to focus only on the breath going in your nose and out of your mouth.

If you're having trouble getting in the present moment, place your hand on your stomach and focus on breathing for five seconds. Feel the rise and fall of your abdomen with each breath.

If your mind wanders, that is normal. Don't be upset or stressed. Instead, gently try to bring it back to focus by using a word such as *breath* or *focus*. It sounds simple, but for many people, it's harder than it looks. If thirty seconds feels like thirty minutes, don't feel bad; it can take some practice. If you can do thirty seconds, be proud of yourself, and tomorrow try to add another thirty seconds. Think of it like going to the gym and working up reps.

Mindfulness does not have to be you sitting on the beach, with the perfect sunset and perfect spandex outfit. You don't have to sit in a room with a group of people, taking deep breaths and repeating mantras. If that's what works for you, that's fantastic, but for some people, the idea of breathing exercises or sitting cross-legged in a room full of people stresses them out. And that is understandable as well.

Here are three simple ways to practice mindfulness that do not involve a breathing exercise: mindful eating, a mindful walk, and a mindful hobby. You can start doing them today.

Mindful Eating

Mindful eating is the idea of taking time at each meal to relish the taste and texture of each bite. It sounds unbelievably simple, but this stress-relieving tactic is also critical in healthy eating programs. It can be so easy to scroll through your phone and not pay attention to what and how much you are eating. When you eat and don't focus on the food, you can miss signals from the stomach to the brain that you are full. Taking a moment to focus on what you're putting into your body is a powerful yet straightforward way to practice mindfulness at each meal or snack, and it comes with the added benefit of stopping you from overeating.

A Mindful Walk

It sounds so simple to pay attention while you walk, but officials are now padding the lampposts on some streets in London because people are walking into them while distracted by their phones! A mindful walk includes taking time to feel your feet on the ground and the wind on your face. A study published in *Frontiers in Psychology* discovered that twenty minutes a day of walking in nature—including a local park or garden if you live in a bustling city—causes stress levels to plummet.[10] As a result, physicians are starting to prescribe "nature" and "green time" to treat stress, anxiety, and depression, as we realize nature can nurture.[11]

A Mindful Hobby

Mindfulness is being in the present moment and enjoying what you are doing. Find activities that you enjoy, and give yourself permission each day to set aside a few minutes to embrace those activities, without distraction, to give the stress response system in your brain a break. It's one of the most powerful things you can do for your brain and body and a too-easily-forgotten component of your health. If you're playing golf, are focused on what you're doing, and you're in a positive mindset, that's mindfulness. If you're playing golf and you're throwing your clubs because you sliced a shot, that's not.

Other Ways to Manage Stress and Cultivate a Positive Outlook

Beyond mindfulness, other insights into how the brain works can help us manage stress and find positivity.

Perspective: What Do You See?

Maybe you've seen one of those optical illusions that can appear to be a rabbit or a duck depending on how you look at it. Two people can look at the same image and see different things. Stress works the same way. Two people might experience the same stressful situation but have very different responses. Stress can actually be quite useful—and you can train your brain to view stress in exactly that light (just as you can learn to see *both* a rabbit and a duck in the drawing).[12]

Even a bit of stress from sitting in traffic or standing in line at the grocery store can be beneficial.[13] These bursts or moments of acute stress can be healthy for the brain for a variety of reasons; for instance, a quick dose of cortisol can turn on the process of autophagy, which cleans up waste and toxins. So next time you're in the "10 items or less" lane at the grocery store and the person in front of you has 37 cans of cat food and 37 different coupons, don't curse them. Thank them—and by doing so, shift your perspective on stress as something that can be beneficial.

Think of a professional versus an amateur athlete. At the moment before they have to play, both can have a similar heart rate and blood pressure. The difference is that the professional athlete will say, "I'm pumped and ready." The amateur will say, "I'm freaking out and overwhelmed." Professional athletes have often been trained to have a perspective where they use stress as energy.

Handling stress is not about *suppressing* emotions. A study published in *Emotion* found that suppressing feelings was much less beneficial to individuals than acknowledging stress and looking at that stress as energy.[14] For example, tell yourself, "This stress is natural and normal and gets my heart pumping so that I can take action on this one task (not two or three tasks), and then I'll get to take a real break."

Schedule Time to Worry

There is nothing wrong with worrying per se; it's when it takes up too much of your time and mental energy that it's a concern. If you find yourself spending a lot of time worrying, here's a silly sounding but very effective trick: schedule time for it. That's right, set aside time, say from 4:00 to 4:15 PM, and don't think about anything else. When the time is up, stop, and move on.

We can take a couple of lessons from professional baseball players, who must master dealing with stress in order to perform. One of their sayings is,

"Turn the page." If something is gone, don't let it interfere with the present moment. (In fact, baseball is filled with mindfulness.) They also say, "Take one pitch at a time" and, "Don't face an opponent you aren't playing right now." That means don't spend your time worrying about things that may happen in the future. Do the same thing with your scheduled worry-time: acknowledge your worries about the past or the future, turn the page, and let them go.

Acknowledge the Negative, Be Thankful for the Positive

It's easy to say, "Just stay in the present moment," but negative thoughts do seem to creep into our brains. Psychologist Rick Hanson suggests that if people give you nine compliments and one criticism today, you're going to be focusing on the one criticism and ignoring the nine compliments when you're lying in bed tonight.

Now, sometimes negative thoughts have their place and we can learn and improve from them. The issue is our brains have a strong tendency to fixate on negative thoughts, due to the fight-or-flight instinct we inherited from our ancestors (and which was, and is, critical for our survival). Acknowledging that fact is the first step in stress management. How did we develop this tendency? Take a moment and imagine yourself with your ancestors a long time ago. You peer out of a cave and you see a tiger. What do you imagine is going to happen next? If your ancestors pictured themselves frolicking with the tiger, maybe going for a carefree ride on its back, they likely didn't live to pass on their genes. We are the offspring of those stressed-out, negative thinkers who thought about the worst case scenario; in terms of evolution, happiness wasn't nearly as important as survival. I'm not saying that being aware of real danger (e.g., a tiger) is the same as fixating on negative thoughts. But I am saying there is a reason that our brains will prioritize and focus on things that are frightening, stressful, or unpleasant.

The second step in stress management is to accept that the brain focuses on the negative. We call this the negativity bias.[15] There is nothing wrong with negative thoughts. The problem is when we dwell on them and don't let go. These negative thoughts can eventually lead to more stress, and sometimes, that becomes the type of stress that can feel unmanageable.

The third step is to accept that the brain tends to think about things it doesn't have. Your ancestors focused on what they didn't have, innovated, and solved problems. This can be helpful, but it can also be a trap: Let's say that I'm going to take you for ice cream, and I tell you that you can pick any flavor you want. It's pretty likely that while you're eating your ice cream your brain starts to think, "Maybe I should have tried the mint chocolate chip instead," "I wonder what the double fudge rainbow would have been like," or, "Maybe I should have gotten the pralines and cream today." Our brains naturally think about all the things we don't have, and we ignore that thing we're holding right there in our hands.

How do we switch our natural inclination from negative to positive thoughts? One way: practice gratitude. Gratitude practices help the brain see positive items to focus on and rebalance the negative bias. A study published in *Review of Communication* found a positive link between gratitude and well-being.[16] To flip the focus from negative, stressful thoughts to positive thoughts, try this:

1. Draw a line down the center of a piece of paper.
2. On one half of the page, write down everything you are stressed and angry about.
3. On the other side of the page, write down everything you are thankful for.

If you list the people you care for and things you are thankful for, that list can get quite long. Don't forget to include the big things like the sun that came up today, the air we can breathe, and all the little things we can take for granted. By having this list right in front of us, we can rebalance the brain to focus more on the positive aspects we tend to minimize and let go of the negative aspects that we tend to fixate on.

Happiness Boosters

Be here now. Here's a quick one-minute mindfulness exercise you can do anywhere to be in the present moment and reset your stress levels: Put your phone away. Name three things you *see*. Name three things you *smell*. Name three things you *hear*. Now breathe in and out deeply three times.

Picture this. Participants in a recent study at Massachusetts General Hospital were asked to find a photo that reminded them of a great memory. Once they found the photo, they were instructed to write one sentence summarizing the memory. This simple practice caused a significant happiness boost that lasted twenty-four hours.

Rose. Thorn. Bud. Put things into perspective. Take a moment and think of the best thing that happened from the last twenty-four hours. That's your rose. Next, think of the most challenging part of the previous twenty-four hours. That's your thorn. Now think of something specific you are looking forward to in the next twenty-four hours. That's your bud. This technique has been shown, in a 2019 study, to be an effective means to manage stress and boost happiness.[17]

Back to baseball. I was watching an interview with a pitcher named Joe Musgrove after he threw a no-hitter, a rare feat. In fact, the team he was pitching for, the San Diego Padres, had never had a no-hitter. Musgrove discussed how in the last few innings of the game, as he realized he was getting close to throwing a no-hitter, negative or anxious thoughts such as "don't blow this!" and "my dad is watching and I don't want to disappoint him" made their way into his brain. Musgrove talked about how he focused on his breathing and the immediate next pitch to push these thoughts out of his head.

We might think that exceptional performances are void of negative thoughts, but most people are not functioning without a battle inside their brains, even at the highest levels of performance. Needing a moment-to-moment practice to focus the brain on the positive and having negative thoughts is normal. We cannot control much of the stress surrounding us, but day by day, we can build connections in our brains to be closer to being experts in happiness and managing stress.

CHAPTER 12

You've Got to Move It

EXERCISE IS LIKE ANOTHER MIRACLE DRUG FOR THE BRAIN.[1] WHILE you're not going to show off your hippocampus in a tank top, if a drug generated the brain benefits that exercise produces, there would be lines stretching miles long to obtain it. That's why in this chapter, we'll look first at how physical activity helps mental activity, then I'll share insights into making an exercise routine that sticks so you don't end up with another Shake Weight or ab roller collecting dust in the closet.

Part of what makes exercise so good for the brain is its impact on the immune system. A 2019 analysis confirmed that exercise is critical for a healthy immune system.[2] Immune cells like T cells are like airplanes; they work best when they are constantly flying. If they don't circulate, they break down, leaving you more vulnerable to viruses and other infections. A consistent exercise routine keeps those T cells in motion and lowers the chances of getting everything from the sniffles of the common cold to the flu.

Another key aspect of why exercise makes such a difference in brain health is the impact that it has on your heart. We know that exercise that gets the blood flowing lowers blood pressure and improves heart health, both vital for day-to-day brain health and preventing depression and dementia. A review of multiple studies found that a workout enhances metabolism, regulates hormones, and balances neurochemicals.[3] That's without even adding in the fact that exercise can increase the size of your brain; cardio workouts that get your heart rate up pump up the volume of gray matter in the brain. Healthy gray matter is critical for memory.[4]

There is a mountain of evidence that exercise makes your brain cells communicate with each other more effectively at any age.[5] That increased brain communication improves mood, boosts happiness, and can even make you score better on a test.[6] Let's take a look at what exercise does for the aging brain. A study published in the *Journal of Alzheimer's Disease* investigated two groups of adults sixty and older. One group did a year of aerobic exercise, and the other group just stretched. The group that did aerobic exercise showed a whopping 47 percent average increase in memory scores after that year, while the stretching group didn't see an improvement in memory. Brain scans of the exercising group showed a significant increase in blood flow to two critical areas involved in storing and retrieving memory: the anterior cingulate cortex and the hippocampus. Participants who exercised displayed a slower breakdown of the hippocampus, while those who scored lower on fitness displayed a faster deterioration of brain cells.[7]

These types of results were replicated in a startling study published in *Neurology*.[8] The study asked fifty-year-old women to work out on an exercise bike, then placed them in four categories depending on how long they could work out. The categories ranged from "high physical fitness" (most fit) to "could not finish the fitness test." After four decades, the researchers found:

- 5 percent of the women in the high physical fitness category developed dementia.
- 25 percent of the women in the moderate and low fitness categories developed dementia.
- 45 percent of women who couldn't finish the fitness test developed dementia.

That means that women who were classified as highly physically fit at fifty years old were *90 percent less likely* to develop dementia than the group that could not finish the test, decades later. What's more, the 5 percent in the high physical fitness group who did develop dementia developed it, on average, at age ninety versus age seventy-nine—the average for the other groups.

Another study looked at a group of men and women with an average age of sixty-six and found that those who worked out between two and five times a week significantly increased the size of the left region of the hippocampus.[9] The left side of the hippocampus is involved in consolidating or strengthening semantic memory—which is involved in word and number recall and understanding language. Any aerobic exercise did the trick, including stationary cycling, walking, and running on a treadmill.

This is just a sample of key studies that highlight how good exercise is for the brain.

So How Much Do I Need to Exercise?

Here comes the big question: How much exercise do you need to do to protect your brain and keep it plump? Do you need to drop everything and train for a triathlon? Fortunately, no. It turns out even small changes can have a big impact.

One of my favorite photos is of people taking the escalator instead of the stairs to the gym. I admit that I have done this myself on occasion, but the next time you are given a choice between stairs and an elevator or escalator, consider this; a 2016 study found that people aged nineteen to seventy-nine who consistently took the stairs rather than an escalator or elevator had a younger-looking brain.[10]

We can easily think of exercise as something that has a long-term protective impact on brain health—but it would be a mistake to think that's all it does. While exercising can feel like saving for retirement instead of enjoying a splurge purchase now, the truth is that even a single exercise session makes a difference. One study found that a onetime ten-minute workout boosted brain regions involved in focus and problem-solving, including the hippocampus.

Exercisers also scored higher on a memory test after just a single workout.[11] In other words, a quick workout can boost your brainpower, and you reap the benefits immediately. So, if you want to ace that test, up your game at work, or just remember more, don't forget to get moving right before. A 2022 study also found that those with depression showed an improvement in mood immediately following one bout of a cycling workout.[12] Furthermore, just one hour of exercise a week has been found to lower the risk of repeated episodes of depression for those with major depressive disorder.[13]

Keep in mind that there can be such a thing as too much exercise. We are—thankfully!—rethinking the exercise mantra "No pain, no gain." Too much pain equals too much stress, and too much stress equals inflammation. The right amount of exercise is the amount that energizes, does not exhaust, and leaves you with enough energy that you want to exercise the next day.

Multiple studies suggest that a great goal is about 120 minutes a week of moderate exercise for brain health.[14] If your goal is to work out five days a week, that means 24 minutes each day. A moderate workout is one where you can talk during the workout, but you can't sing (which is good if you are working out with me, because you don't want to hear me sing).

Does It Matter if It's High- or Low-Intensity Exercise?

An intense workout is exertion that leads to about 70 to 85 percent of your max heart rate according to the American Heart Association. That means you couldn't talk during the workout. Some examples are running or biking at ten miles per hour or faster. A low-intensity workout is anything else.

When people engage in vigorous exercise, they increase two essential neurotransmitters: glutamate and gamma-aminobutyric acid, or GABA, which help your brain cells communicate.[15] That's why high-intensity exercise is an essential part of treatment for mental health, specifically depression. People with depression can have fewer types of specific neurotransmitters and key chemicals, but there is evidence that exercise can provide a boost of neurotransmitters and key chemicals such as BDNF, a growth factor for brain cells.[16] The brain needs to keep the body moving, which is more challenging than studying for any exam, and the accompanying boost in neurotransmitters then helps the brain work more effectively in moments when we are not exercising.

Intense workouts can also help manage conditions like diabetes. A 2017 study found that just two weeks of high-intensity interval training (a HIIT workout) could improve glucose metabolism in muscles and insulin sensitivity in those with type 2 diabetes.[17]

But adding some intensity to your workout doesn't mean you need to feel like you need CPR. Even if you can't do more intense workouts, just staying active and moving throughout the day is greatly beneficial to the brain.

Walking: An Ideal and Accessible Form of Movement

A study at Cardiff University in the UK followed thousands of people for over thirty years and found that simply walking thirty minutes a day lowered the risk of dementia by about 65 percent.[18] The best part? Those thirty minutes didn't even have to be done consecutively. Why is walking so effective? It's likely our ancestors walked long distances searching for food. Some of them found their way back home and remembered where those sources of food were. Those who couldn't remember where to find good food sources (or got lost on the way back) likely became food themselves and didn't pass on their genes. We are the offspring of the people who walked and remembered.[19]

Another study investigated which factors were associated with reaching your one hundredth birthday. Right at the top of the list was living in a walkable neighborhood.[20]

How Much Should You Walk?

You may have heard that you should be aiming for 10,000 steps a day. But that number doesn't come from a scientific study. The source is a marketing agency in Japan looking to sell pedometers, and they came up with the 10,000-step goal.

So, is there *any* evidence for getting 10,000 steps? A study published in *JAMA* stated, "Compared with taking 4,000 steps per day, taking 8,000 steps per day was associated with significantly lower all-cause mortality."[21] Meanwhile, another study from *JAMA Internal Medicine* stated, "Among older women, as few as approximately 4,400 steps/d [per day] was significantly related to lower mortality rates compared with approximately 2,700 steps/d. With more steps per day, mortality rates progressively decreased before leveling at approximately 7,500 steps/d."[22] In other words, there is nothing wrong with more than 7,500 steps, but you might not be getting much extra health benefit. Tracking your steps can be a great motivator, but you can also just consider taking a forty-minute walk three times a week. A study found that people who followed this routine increased the size of their hippocampus. Not only did these walkers see an improvement in memory but their brains also looked about a year or two younger than their brain's chronological age.[23]

When Should You Walk?

Go for a walk *after* you eat. Also, go for a walk *before* you eat. Why? Take a moment and think about our ancestors. If they could see us now, making a couple of swipes on our phone and food appearing at our doorsteps soon after, their heads would explode. Even a couple of generations ago, obtaining and preparing food took time and effort. The goal of the day was to get food, prepare food, and repeat. We had to *physically move* to get our food.

Walking within a half hour before eating can lower the amount of fat and sugar in the blood after eating.[24] When we move our bodies after eating, we help regulate metabolism as it helps move sugar into the muscle.[25] Remember, unmanaged diabetes is one of our single most significant risk factors for developing dementia, but exercise has a substantial and undeniable impact on the management of diabetes. A study found that if older people go for a fifteen-minute walk—no HIIT-level intensity necessary—after each meal, it could help regulate blood sugar levels and stave off type 2 diabetes.[26] The next time you eat something, ask yourself how much effort you put into getting the food. Making time for a quick stroll around the block or a little walk up and down a flight of stairs could help your brain.

Make It a Habit

Exercise every day—or at least don't go more than two days in a row without exercising. Consistent exercise enhances insulin action, which affects whether or not an individual has diabetes. Furthermore, adults with type 2 diabetes should take care to diversify their workouts. They should be a mix of aerobic, balance, and strength exercise, totaling about 120 to 150 minutes a week.[27]

Does Walking Really Help Your Brain?

Vasaloppet is the oldest cross-country skiing race in the world. It takes place in Sweden and was inspired by King Gustav Vasa's journey in 1520 when he fled from enemy soldiers. Today, about 10,000 to 20,000 people participate in this annual event. A group of scientists decided that since about 200,000 people had participated in the race over the span of ten years, it would be a great way to study the impact of exercise on heart and brain health. After collecting and analyzing their data, researchers found that cross-country skiers had a lower incidence of heart attack and vascular dementia. In addition, they were half as likely to be diagnosed with depression, and they had a delayed manifestation of Parkinson's.[28]

So, should we all grab some cross-country skis and make a trek to Sweden? Although a trip to Sweden sounds great, the scientists involved in the study suggested that it's the action of feet hitting the ground during cross-country skiing that protects the heart and brain—which is similar to what happens when we walk.[29] When our feet hit the ground, a pressure wave is sent up our legs and tells our heart the proper amount of blood to send to the brain.[30]

A study published in the *British Journal of Sports Medicine* looked at women aged seventy to eighty who were living independently.[31] The researchers took pictures of the participants' hippocampi and tested their memory before and after they tried six months of balance exercises, two weekly hourlong walks, or weight training. Only walking showed an impact on the size of the hippocampus. Weight training and balance exercises are also excellent for the brain, but if you are able, it is essential to include walking for the health of the hippocampus and boosting memory.

When You Train: Working the "Go" Muscles and the "Show" Muscles

When you're exercising, don't forget to work out the "go" muscles as well as the "show" muscles. The "show" muscles are the biceps and pecs, the ones we flex and look at in the mirror. The "go" muscles are the leg and back muscles, the ones that help us get from place to place. Even though it's not uncommon for gymgoers to skip leg workouts, that's a mistake—research tells us that muscles in the legs and back are particularly important for our brain function.

Some of the evidence for this connection comes from people who unfortunately lose function in these muscle groups. Patients who cannot walk due to multiple sclerosis and spinal muscular atrophy often experience rapid mental decline. While scientists knew this was true, it took a groundbreaking study, published in 2018 in *Frontiers in Neuroscience*, to uncover why. It turns out that when people cannot do load-bearing exercises with their legs, it negatively affects brain chemistry. Weight-bearing exercise sends signals vital for producing healthy neural cells, and these new nerve cells help the brain adapt to stress and slow down the brain's aging process.[32]

Other evidence for the importance of "go" muscles comes from space. Astronauts who experience prolonged periods of weightlessness see negative impacts on their immune systems. Why? Part of the issue is the lack of weight-bearing exercise. However, a consistent routine of weight-bearing exercises can rebalance the immune system.[33] If you've seen images of an astronaut on the space station exercising, it's not just because they miss the gym. They are also protecting their brains and immune systems.

Good News for Yoginis

A study published in *Brain Plasticity* found that after eight weeks of yoga, the brain can better handle stress. Furthermore, yoga practitioners improved their scores on tests of decision-making and attention.[34]

Making Exercise a Habit: CARS

At this point, you might be saying, "I get it! Exercise is good for my brain. But if the key to any exercise plan is consistency, how do I stay with it? How do I make exercise a habit instead of that exercise bike or treadmill becoming a very expensive coatrack?"

I have a family member who struggles with health issues, including obesity and diabetes. This person clearly understands the benefits of exercise but, like millions of other people, has a tough time sticking to an exercise plan. The New Year's resolution always ends the second week of January. Recently I helped them maintain an exercise routine by employing some brain-science-based habit-forming techniques. Here is the trick: there are four critical steps to form a habit:

1. Cue
2. Action
3. Reward
4. Support

Cue

Most of your day—95 percent of it, in fact—is spent in an unconscious state. Don't believe it? How many times have you left your home and wondered if you closed the front door or the garage door? Or whether you left the stove on? When we do repetitive actions, we are not paying attention; it is like your brain is on autopilot. So, when it comes to making exercise a habit, we want to incorporate exercise into our "default" or autopilot mode—this makes it easy for us. Now, I am not saying we need to exercise on autopilot like zombies! Instead, we can set up a cue that drives us to take action to stick to an exercise program. This cue system is similar to how we set up a cue to help us fall asleep or remember to floss.

Here are two cues that can be helpful: One visual cue is to simply place your walking shoes where you can't miss seeing them, as a reminder to take your daily walk. Another visual cue can be a smartwatch that reminds you of your steps.

One of the most powerful methods for creating a consistent exercise routine is a trick called habit stacking. Pick a habit that you already have and make it the cue for your new habit. For example, most people brush their teeth, but they do not floss. Research into habits has shown that when people place their floss right next to their toothbrush, they develop a flossing habit.[35] You can do the same thing with exercise. Habit-stack your morning coffee or favorite daily TV show with some simple exercise, or engage in your new habit right before your daily habit. And if you want to combine visual cues, put your floss in your shoes by the front door. (OK, don't do that one.)

Action

To start, keep the action—in this case, exercise—simple. Have you ever tried to make too many changes at once and end up changing nothing? When it comes to exercise, it's best to start small and build.

For example:

- Walk in place during TV commercials or before the next episode of what you are streaming.
- Stand and/or walk during phone calls.
- Park farther away from the grocery store.
- Do ten jumping jacks every hour (you can set your smartphone to remind you to do this).

Reward

Exercise has to be fun and rewarding, or your brain will likely not make it a habit. This is why pairing the exercise habit with something you already love doing can be very helpful. Have you heard someone say, "I walk on the treadmill every single day, and I hate every single second of it"? If you enjoy working out on a treadmill, that's great, but if it's drudgery, ditch it and look for a new and enjoyable experience, as newness and novel environments are keys to focus and memory. Try hiking new trails by a lake or just walking in a new neighborhood. Learn a new sport and have fun doing it.

The body can acclimate to the same activity, and results diminish. The same is true with the brain. Try to keep your workouts fun and fresh. There are endless virtual workouts online and classes at local libraries that teach yoga, dance, Pilates, and tai chi.

Support

A study found that active participation on a sports team had brain health benefits.[36] It wasn't just the playing but also the socializing that boosted the brain. Find a workout buddy or join a sports team or exercise class.

Penn Jillette of the magic duo Penn and Teller lost a hundred pounds. To keep himself motivated, he used a brain trick that would terrify most people. At the beginning of his weight loss journey, he posted his weight on social media every day. That may sound horrifying to you, but that accountability kept him motivated. Similarly, my family member texted each morning right after taking a ten-minute walk. That little bit of accountability was critical to keeping at it.

Be aware, though, you're going to have good days and bad days when you're establishing your exercise habit. Stick with it! If a forty-minute walk is too much to fit in, even three short (think ten- to fifteen-minute) walks a day can reap significant brain benefits. Each day you stick to the habit, you are building and strengthening the connections between brain cells that the habit establishes in your brain. Through repetition, those connections will get stronger and stronger, and a healthy habit will be formed. And remember there is overwhelming evidence that exercise not only slows down and reverses brain age but it can also improve the way your brain works each day.[37] The secret to a healthy brain is just a couple (or few thousand) steps away each day. Walk to your favorite places—and, beyond getting support, there's a brain bonus if you can do it with your favorite people. In the next chapter, we'll find out why.

CHAPTER 13

Get Yourself Connected

CAN GOING TO A DINNER PARTY PRESERVE YOUR MEMORY? PEOPLE over fifty-five who regularly participated in, or hosted, dinner parties or other social events had a lower risk of losing their memories.[1] It wasn't because of what they ate or where they went; it was the effect of the repeated social connection with other people.

How important is social interaction to our mental and physical health? We are used to hearing about smoking and obesity as risk factors to our wellness, but loneliness has the same impact on mortality as smoking fifteen cigarettes a day, making it worse for us than obesity.[2] Loneliness increases the risk of heart disease, Alzheimer's disease, depression, anxiety, high blood pressure, stroke, and even premature death (before the age of seventy-five). For older adults, loneliness can increase the risk of early death by 14 percent. Loneliness increases the risk of dementia by 40 percent.[3] We also now know that loneliness is associated with poorer decision-making, attention, cognitive ability, and brain shrinking and aging.[4]

It's a relief that having fun and hanging with friends protects your brain.

A couple quick and important points. It is normal to feel lonely from time to time, but there is a greater risk of loneliness as we age, as family and friends pass and children move away—that kind of lasting loneliness that makes us feel socially isolated. Half of adults feel lonely.[5] Around one-third of adults say they lack regular companionship.[6] But engaging in meaningful relationships boosts well-being.

A meaningful relationship is a relationship where individuals feel supported during good times and difficult times. During adversity, a meaningful relationship helps buffer negative effects of stress and helps the individuals thrive. During good times, a meaningful relationship helps support personal growth by encouraging people to embrace new opportunities. A meaningful relationship is also based on trust and being heard.

Everyone can create and cement meaningful, long-lasting relationships. It's not just about surrounding yourself with more people—it's about finding the right types of relationships, as well as some nonintuitive factors.

The Impact of a Lonely Brain (or, What We've Learned from Astronauts)

Before we dive into ways to combat loneliness, let's take a step back to examine a critical question: Why does loneliness have such a striking impact on mental and physical health?

Loneliness turns on genes that increase inflammation, elevating the risk of conditions from heart disease to depression to dementia.[7] Loneliness leads to an increase in the stress and fight-or-flight responses. This suggests that an increase in cortisol turns on inflammatory genes that lead to inflammation.

A factor in feeling lonely is isolation. Of course, one can be isolated and not lonely, but isolation can increase the risk of feeling lonely. Scientists at the University of Arizona found a clever way to study the effect of isolation on the immune system. They turned to a job that can get very isolated: astronaut. Surprisingly, though astronauts are not exposed to many other people in space, they still become ill from viruses. These viruses are not in their environment; they're lurking inside the astronauts' bodies, where they are usually kept in check by the immune system. Researchers suspected isolation suppressed the astronauts' immune systems and allowed the viruses to become active. In order to get a larger sample of subjects than just astronauts, researchers replicated the isolation aspect of being in space by studying individuals who worked in Antarctica.[8] Further studies found that if isolated individuals ate a Mediterranean-style diet, managed their stress through techniques such as mindfulness, and performed short bursts of high-intensity exercise, their immune systems were no longer suppressed.[9]

What Makes Us Feel Lonely?

A region of your prefrontal cortex called the medial prefrontal cortex holds a map of your social networks.[10] When you think about someone, this part of your brain shows a specific kind of activity based on how close the person is to you. For example, the activity will be different if you think about a close friend versus an acquaintance. Lonely individuals can experience difficulty feeling closeness.[11] They tend to have the brain activity associated with thinking about an acquaintance when they think about someone close to them. That's why just being alone doesn't mean someone feels lonely, and why an individual can be surrounded by others and still feel lonely. Loneliness isn't related to how many people you know but instead how close you feel to them. It's the lack of closeness that raises the risk for inflammation.[12]

We can think of loneliness in the same way we understand physical pain. Physical pain alerts us that we need to take care of our bodies and heal. Loneliness is a signal that we need social interaction for the health of our brains.

A significant study found that feeling lonely was 14 to 27 percent based on genetics.[13] While genetics can predispose people to loneliness, environment plays a more significant part. Loneliness can be caused by such factors as physical isolation resulting from relocation, divorce, or the death of a loved one. It can also be caused by low self-esteem and not feeling heard or understood.

How to Combat Loneliness

Feeling lonely is not irreversible. While it's difficult to escape the hopelessness that often accompanies loneliness, there are things that can help you fight it off.

You Can Use Tech to Connect—But Keep Your Use in Check

Should we turn to social media and Facebook friends and followers to alleviate feelings of loneliness? It's not a simple yes or no. On the one hand, technology can be a helpful tool to alleviate loneliness. For example, a 2021 study found that chatting online with friends and family or through multiplayer games helped teens find support and feel less lonely.[14]

On the other hand, prior research found that interacting on Facebook, Snapchat, and Instagram *increased* feelings of loneliness. When participants in a 2018 study spent less time on social media, they demonstrated fewer symptoms of depression and loneliness.[15] The key is how the tech is being used. At a time when we're more connected than ever via our phones and other devices, we may also be more isolated than ever as real-life social connections are replaced by virtual connections online.

Data suggest that social media use can increase a commonly described feeling of FOMO, or fear of missing out. Interestingly, FOMO does not apply to everyone on social media, but those who are prone to feelings of missing out can have those feelings exacerbated.[16] It can seem like everyone on social media is riding a jet ski while drinking champagne. Social media also gets people to focus on everything they don't have instead of all they have. The grass tends to look greener in other people's Instagram posts (and it's not just because they used a cool filter).

There's also the fact that our brains love two things: surprise and reward. When we are surprised or rewarded, our brains release feel-good brain chemicals such as dopamine. What does the brain find surprising and rewarding? New information. It doesn't matter if it's good news or bad news; the brain craves new information. Why? Looking back to our ancestors again, those who wanted to learn new things had a survival advantage. They figured out new and better ways to gather food and protect themselves when they paid attention to new information in their environments. Cell phones, tablets, apps, and websites are designed to hijack our brains and make these devices hard to put down. Notice how our phones are full of unexpected beeps and buzzes that surprise the brain with the promise of new information. The combination of surprise (you weren't expecting that push notification) and reward (now you know something you didn't before) can make the brain mesmerized and fixated.

Remember the days when you got your news from the morning paper and the evening news on TV? Today we live in a world where breaking news interrupts breaking news, and much of it is bad news. With each story about bad news, our bodies release cortisol. The more bad news, the more cortisol. Extensive exposure to negative news stories increases stress.[17] And guess what: this constant state of stress is a risk factor for feeling lonely, because in some cases, stress can make an individual withdraw from others.

But you don't need to throw your cell phone off a cliff or go cold turkey on social media and screens. To avoid doom-scrolling, schedule your news and social media time for specific fifteen-minute sessions during the day. Put them on the calendar. To get the best out of technology, consider the quality of online interactions and consider limiting your time on devices and social media and including in-person interaction and connection. There are times when video and phone calls can be very helpful to mental health. When used in moderation, apps like Zoom and FaceTime can bring us closer together. Online courses can be engaging and beneficial. A study of people ages 52 to 104 found that participating in an online exercise course decreased loneliness and social isolation.[18] Remember, it is not necessarily the use of screens so much as how they are used that matters.

The bottom line is that technology and social media are not all good or all bad. They absolutely can be used in moderation and have brain health benefits. It is just important to be aware that there are negatives to these technologies as well. One final point: Try not to let social media and technology be your only means of social interaction; make it a point to find other ways to engage in person or without screens.

Fun with Friends . . . and Even Family

A study investigated whether people find more joy being with their friends or family. (In this study, *family* was defined as people who live in one's home.) They found people were happier when they spent time with their friends instead of family.[19] Ouch to our families.

One reason people didn't have feelings of happiness and joy around their family members was that the types of activities they did together at home were more likely to involve chores: doing laundry, washing dishes, taking out the trash, housework, paying bills. Not exactly New Year's Eve!

We don't usually say to our friends, "Can you come over and go over my credit card statement? And after that, how about we sort through my junk mail?" We tend to reserve fun activities for our friends. We need to make an effort to ensure we don't make it all about chores and to-do lists with family.

You might also be saying, "But my friends and family cause me stress! Aren't I better off without that drama?" Of course, there are toxic relationships that need to be avoided, but even healthy relationships have good and bad elements. If we aim for perfection in our relationships, we can end up constantly disappointed. Miscommunication, misunderstandings, or disagreements can sour a relationship. So how do we maintain a healthy relationship with the people we love and care about even when they have inevitable bad moments? Well, science has an answer for that. Researchers assembled a group of married couples and divided them into two groups.[20] They asked the couples in the first group to tell the story of how they met. They asked the second group the same question, but with a twist: don't just tell the story of how you met, describe how you almost didn't meet. They heard: "Every day I went to the third floor, but that day I happened to go to the fifth floor." "I always turn right at that corner. That day I turned left."

Next, they gave all the couples a quiz on marriage satisfaction—and the people who told the story of how they almost didn't meet scored higher. When this group forced their brains to think of all the ways their meeting could not have happened, they pushed their brains to unadapt.

The concept of unadapting in the brain is based on the idea that the brain tends to adapt to the good or positive and fixate on the negative or bad. If something good happens to us, we easily adapt and move past it. The celebration of a promotion or goal achieved can be fleeting, lasting seconds to hours, but we can harp on a mistake for decades. Unadapting is a practice that allows the brain to focus on the positive. The idea behind the study was that we can easily forget the miracle that went into just meeting someone special. The storytelling technique encourages the brain to have gratitude, and it increases feelings of happiness and satisfaction.

Caregivers Take Care!

When flight attendants go over the safety protocol on a plane, they announce that in the event of an emergency, you should place your oxygen mask over your face before placing the mask over a child or someone who needs help. I remember being a kid and thinking this was a bit harsh.

Anyone who takes on the role of family caregiver, whether for children, or aging or ill relatives, or both, takes on the unseen labor of caring and ensuring the household runs smoothly.

The caregiver mentality is admirable, but there can be negative consequences when caregivers neglect their own health and emotional needs. Compared to the general population, people who are caring for a spouse with dementia are at a higher risk of developing the disease themselves. Why? Because of chronic stress.[21] The chronic stress response causes the brain to shrink, damaging the hippocampus. We want to care for the people we love—and we should—but we also must take care to pay attention to our own health. So, if you're a caregiver, ask for help when you need it, don't take the world on your shoulders, and make sure to do the self-care practices covered in this book to combat the impact of chronic stress. Take a cue from airplane safety: If you don't take care of yourself, you won't be able to take care of someone else.

Train the Brain for Resilience

If we can consciously create feelings of happiness and satisfaction, can we consciously let go of anger and resentment? Others may say or do things that intentionally or unintentionally hurt our feelings. We may ruminate on these negative thoughts and interactions, which releases cortisol. As we have seen, when our brains are chronically bathed in cortisol, this increases the stress response and can damage the brain. It's no surprise that people who are better at adapting to stressful situations and releasing their anger are much less likely to be lonely—no matter what their age.[22] A resilient brain is protected against stress-induced changes to its structure and function. For example, as we discussed earlier, increased cortisol release can shrink the hippocampus. A resilient brain is protected from this change even in the presence of high levels of stress.

You might say, "I am aware I should let go of these negative feelings, but how can I actually do it? I can exercise and eat right to build a resilient body, but how do I build mental resilience?"

The brain can be trained to be resilient using a fifteen-minute practice. Before I tell you how to do it, let me tell you about the Chicago Cubs. The Cubs went 108 years without winning a World Series. That's a lot of stress and disappointment for the players. The Cubs players started doing this fifteen-minute practice in the locker room before games, and the following year they won the World Series. OK, I am not saying that is the only reason. But if baseball isn't enough to convince you, consider this: The US military has found that servicepeople who do this exercise can reduce their stress and find greater resilience.[23] Nurses and doctors at Emory University hospitals are also using this practice to help patients undergoing treatments for PTSD and highly stressful therapies for conditions such as cancer.[24] What's more, this practice has a visible, measurable effect on the brain. One study placed two groups in high-stress situations, and then scanned the participants' brains. The group that did this practice had more connectivity across brain regions, meaning parts of the brain are better able to communicate, making them more resilient.[25]

I hope you're convinced by now. This practice, called self-compassion and acceptance, really works, and it's being used in the real world. Here's how to do it in three simple steps:

1. Think of a person (or a pet) that you love. Sit with that feeling for about five minutes.
2. Take that feeling of love for someone else and give it to *yourself* for about five minutes.
3. Take that feeling of love and send it out to people you are upset with. Even if you have every right to be upset with them, send this feeling for about five minutes.

You might be wondering why a practice called *self*-compassion includes step 3. Why should you send good feelings to someone you're upset with when you have every right to be angry at them? Well, have you ever heard the expression "Forgiveness is for you, not them"? That statement has real credence in brain science. A constant state of anger can damage the brain due to the chronic release of hormones such as cortisol. It is important to take a moment and not only forgive yourself but also forgive others if you are holding on to anger. Of the three steps, though, researchers found that step 2 was the most likely to be skipped. Participants said they could cultivate a feeling of love and send it to someone who had hurt them or with whom they were angry, but when it came to giving themselves the feeling of love, the responses ranged from uncomfortable to *ew, gross*. We tend to focus on everything we do wrong and not what we are doing right; we need to be kinder to ourselves. Vitally, though, those who skipped step 2 did not show as resilient a brain in the scans.

Boost Wisdom

Wisdom has also been shown to protect against loneliness. If that idea sounds a bit abstract, in brain science, *wisdom* is often defined as empathy, compassion, emotional regulation, and self-reflection.[26] Specific brain regions respond to loneliness and wisdom in opposing ways. For example, in recordings of brain activity, an area of the brain called the temporal-parietal junction was more active in lonely people during angry emotions. On the other hand, wiser people had more activity in this part of the brain during happier emotions.[27]

Wisdom and loneliness appear to influence—and be influenced by—gut microbial diversity.[28] This relationship might seem mind-blowing, but it makes sense when we remember the mind-gut connection and that our emotions and mood can impact inflammation, and thus gut bacteria.

We build our wisdom using the fifteen-minute practice and the unadapting storytelling technique detailed earlier in this chapter. Those two simple practices bolster empathy, compassion, emotional regulation, and self-reflection, skills that increase wisdom and effectively counter or prevent severe loneliness.

Find the Purpose

Meaningful pursuits tend to cause a steady release of feel-good chemicals in the brain. People who feel a sense of purpose feel less lonely.[29] A sense of purpose comes from meaningful relationships, mentoring, and feeling like one is working toward a greater good. It might come from parenting, religion, activism, volunteering, or focusing on one's career.

Perform Small Acts of Kindness

To lessen isolation and loneliness, you can boost brain chemicals such as serotonin and endorphins by performing small acts of kindness.[30] One study found that people who were caregivers for a loved one suffering from dementia had lowered feelings of isolation and loneliness when they performed a simple act of kindness toward anyone.[31]

What's a small act of kindness? It might be something like the following:

- Wishing others well or checking in with somebody.
- Giving a compliment without expecting anything in return.
- A phone call to somebody you don't usually reach out to.

While these days many people are uncomfortable with calling rather than texting (and if you feel that way, you are not alone), there is something about the human voice that builds a feeling of closeness in the brain. A 2021 study asked people to reach out to a friend they hadn't heard from in a while. Most participants said they felt awkward making a phone call and would prefer to text. However, not only did the phone call take about the same amount of time as texting or emailing back and forth, but also, when participants made the call, they reported feeling a significantly stronger bond.[32]

Since we know that the human voice builds a feeling of closeness, you may be wondering what happens for those who can't hear. Even if you have strong hearing, our hearing does diminish as we age (especially for men, who are three times likelier to develop hearing loss than women).[33] Let's take a moment to discuss hearing, hearing loss, and how it can affect the aging brain.

The Surprising Link Between Hearing Loss and Memory Loss, Depression, and Anxiety

Studies have shown that, among people who lose their hearing as adults, the more severe the hearing loss, the greater the risk for memory loss, depression, and anxiety.[34] People with mild untreated hearing loss are twice as likely to develop dementia than those without hearing loss. Those with severe loss are five times as likely to develop dementia.[35]

There are a few reasons for this connection between hearing and memory loss. First, hearing loss can lead to social isolation and lack of engagement and learning. Since all three are critical for brain health, their absence can be detrimental. Second, without hearing, there can be a lack of stimulation to brain regions involved in learning, memory, and sound processing. Since the brain works on the principle of "use it or lose it," these parts of the brain can shrink and atrophy. Third, with hearing loss, other parts of the brain involved in learning overcompensate. This can tire these overused brain regions, leading to loss of cognition.

There's also a theory that hearing loss causes changes in brain activity that can promote abnormal proteins that are the hallmark of brain trash.[36] Then, there's the fact that hearing loss can cause a reduction in brain chemicals that are involved in synaptic plasticity, which is change that occurs at synapses, the junctions between nerve cells that allow the cells to communicate. Synaptic plasticity allows the brain to learn new information and adapt to new environments. At the same time, hearing loss can also raise the risk of depression and anxiety by making one feel more isolated and lonelier due to lack of engagement.

It's not just about hearing but also being heard. A study found that in general, people who had someone they felt they could talk to about issues or problems in their lives had younger, more resilient brains.[37]

Do Hearing Aids Help?

Using a hearing aid lowers the risk of developing dementia by 18 percent and depression and anxiety by 11 percent. In addition, those who get a hearing aid are at a 13 percent lower risk of falls, and are thus less likely to get a head injury, too.[38] One study found that 97.3 percent of those who received a hearing aid showed clinically significant improvement in cognitive function.[39] But while mild hearing loss is common after the age of seventy, it is often not diagnosed—and, surprisingly, only 12 percent of those who are diagnosed with hearing loss get a hearing aid.[40]

Using a hearing aid doesn't just improve hearing; it protects the brain, too. If you know someone who needs a hearing aid but is resisting getting one, you can tell them they aren't just improving their hearing; they are protecting their brain.

Other Senses

Taking care of senses beyond hearing can also help combat loneliness and age-proof your brain. Those who need and undergo cataract surgery lower their risk of dementia by 30 percent in the ten years following the surgery.[41]

It is believed that by improving vision, individuals stay engaged and continue to learn new information.

There's also a strong connection between smell and memory. After all, at the top of your nasal cavity is your brain. There's even the chance that we can use smell tests to diagnose dementia and Alzheimer's.[42] People who were not able to identify four out of five common odors had more than double the odds of developing dementia in the next five years.[43] If the part of the brain involved with smell is breaking down and atrophying, it is possible other parts of the brain can be aging as well.

And if you even want to take it up a notch, there is something called "smell training." Smell training involves picking four different types of smells, such as spicy, fruity, flowery, and resinous, and smelling them individually for about ten seconds, twice a day. There is evidence this can help regain sense of smell after infections such as COVID-19.[44] It's like taking your nose to the gym (without having to smell the gym). Bottom line, take time to smell the roses. Smell your food. Not only is it a mindful moment but you're also engaging an important sense. Remember, your brain is use it or lose it. Our senses connect the outside world to our brain—and engaging our senses slows down its aging process.

It's all about connections. Meaningful relationships connect us to other people, and those connections make new connections in our brains.

CHAPTER 14

You Are What You Eat (and So Is Your Brain)

SIXTY TONS OF FOOD PASS THROUGH THE INTESTINES OF AN AVERage human in their lifetime.[1] And if you have ever been on a cruise, you can double that number. (I'm kidding, but I must admit I love cruises in part because I love food.) What we eat can calm or increase inflammation, help us lose or make us gain weight, and even improve our brain function. In fact, diet plays a role in essentially every aspect that we have discussed thus far—our overall brain health, mood, sleep, and productivity, as well as heart, immune, and metabolic health. But could you be eating the right food the wrong way? Or at the wrong times? Let's delve into simple things you can do to feed your brain and surprising new insights to get the most benefits from your next meal.

What's the "Best" Diet for Brain Health?

There are a dizzying number of complex, restrictive, and even bizarre diets out there, and new fads or trends in eating pop up seemingly every week. If we cut through the noise, a lot of the research on the connection between diet and gut/brain/immune health improvements point to a Mediterranean-type diet as being the best. That's a diet filled with fruits and vegetables, bursting with beans, nuts, and whole grains, and featuring fish, seafood, and good fats like olive oil. The Mediterranean diet lowered the risk of Alzheimer's even in those with a genetic risk factor for the disease.[2]

There's also the DASH (Dietary Approaches to Stop Hypertension) diet, which is rich in minimally processed foods, such as vegetables, fruits, fish, olive oil, and beans. A study published in *Alzheimer's and Dementia* found the MIND (Mediterranean-DASH Intervention for Neurodegenerative Delay) diet, a mix of the Mediterranean diet and the heart-healthy DASH, can lower the risk of developing Alzheimer's by about 35 percent—and that's just for people who followed it casually. Those who followed the diet strictly lowered their risk up to 53 percent.[3]

In 2021 the authors of the MIND Diet at Rush University Medical Center released a follow-up study that followed 921 participants who had a mean age of eighty-one for an average of six years. People who regularly included pears, olive oil, tomato sauce, kale, oranges, and moderate amounts of wine in their diets had a 38 percent lower risk of developing Alzheimer's. (Although—to warn you—alcohol is controversial, and we'll talk about that in just a few pages.) Those who *regularly* ate kale, beans, tea, spinach, and broccoli had a *51 percent lower risk* of developing Alzheimer's.[4] Why does this diet work? For one, the foods included in the diet are rich in flavanols, a potent anti-inflammatory that lowers the risk of inflammation that can spread to the brain. The foods in the diet also nourish the 37 trillion bacteria in your gut. Plus, these foods are heart-healthy and benefit the heart-brain connection.

The types of foods we combine in a meal can raise or lower the risk of cognitive decline. For instance, studies have found that those who consume at least one seafood meal a week have a slower rate of memory loss over five years than those who do not.[5] Meanwhile, a study published in *Neurology* followed subjects for twelve years; the results suggest that combining processed meat such as sausage or cured meat with a starchy carbohydrate like a potato or sugary snack increases the risk of dementia.[6] You might be saying, "You're taking away everything that I love to eat!" Don't despair! In that same study, when people ate heart and brain-healthy foods *at the same time as* highly processed/sugary foods, it helped lower their risk for dementia.[7]

A Mediterranean diet is an excellent foundation because it is easier to follow than many other diets. Diets that are too restrictive aren't sustainable in the long term for most people. Instead of thinking about limiting the foods you love, think about *adding* heart- and brain-healthy foods (like fish, walnuts, and olive oil) and eating both at the same time. I am not suggesting you just double the amount of food you're eating—instead, make some key substitutions and additions.

It's OK to Love That Latte

In one study, women sixty-five years or older who consumed more than 261 mg of caffeine per day were also 36 percent less likely to have incidences of dementia over ten years of follow-ups.[8] Now, I'm not saying you need to drink that much—it translates to drinking upwards of two mugs of coffee or five mugs of black tea. That's quite a bit. It's essential to run this by your physician, as caffeine may not be beneficial for individuals with certain underlying conditions.

The Foundations of a Brain-Healthy Diet

Throughout this book, I've referred to the power of our diets to support our health in various ways—limiting LDL (bad) cholesterol, controlling inflammation, helping us get better sleep, and so on. In this chapter, I'm putting it all together so you can assess how you're eating and, if your current habits aren't in line with protecting your brain, you can begin making small changes for the better. I've organized this around what foods (and nutrients) to prioritize and what to limit, and have included a section on supplements. Let's dive in.

Healthy Fats

Your brain needs healthy fats. Brain cells are coated in fats from omega-3. Those fats keep the electrical signals that your brain cells use to communicate flowing from brain cell to brain cell. You need that electricity to travel effectively so your brain works well (and so you'll remember what you read in this chapter when you're deciding between the fried chicken sandwich and the grilled salmon at your next dinner out). You can think of your brain cells like a charging cable: the coating—in this case, healthy fats—keeps the electricity from escaping so it can actually flow from the outlet to your device. But your brain and body cannot manufacture omega-3 on their own. You need to take it in with the right foods. Some key sources of healthy fats are fish, nuts, seeds, and certain oils.

Eating a 3.5-ounce serving of salmon (not fried!) twice a week improved sleep in children and improved scores on IQ tests.[9] That's quite a double benefit! There is robust evidence this is also helpful for adults.[10] Eating oily fish high in omega-3 fatty acids (think in terms of mackerel, herring, lake trout, sardines, albacore tuna, and, again, salmon) is also good for your heart—which, in turn, protects brain health.

Now, while there is strong evidence that eating fish about twice a week is beneficial to brain and heart health, there is a bit of a difference between wild-caught and farm-raised fish. As it turns out, it doesn't just matter what *we* eat. What those animals ate before they made it to our plates also matters. Wild fish tend to be higher in omega-3 because they usually eat more algae and plankton, which is loaded with omega-3. When the fish eat omega-3-rich food, it gets into their tissues, and they pass it on to us.

Farm fish, on the other hand, are often fed corn, which is high in omega-6. Omega-6 has brain and health benefits, but there are more foods that have omega-6 in them, so there is often too much omega-6 in our diets compared to omega-3. Focusing on foods that have omega-3 can help balance the amount of omega-6 and omega-3 that we're consuming. Farmed salmon, though, is still high in omega-3s because of the fat content of the filets.

Fish High in Omega-3 (Wild or Farmed)

- Salmon
- Sardines
- Atlantic mackerel
- Cod
- Herring
- Lake trout
- Canned, light tuna

Other Foods High in Omega-3

- Flax seeds
- Chia seeds
- Avocados
- Walnuts
- Brussels sprouts
- Flaxseed oil (not recommended for cooking but can be used as a dressing)
- Olive oil

Flaxseed and olive oil are other sources of healthy fat. Keep in mind, not all olive oil is created equal. Cold-pressed, extra virgin, minimally processed oil with no additives is best. Look for a date when the oil was bottled or the best-by date. The less time between pressing and ingesting, the better it is for your heart and brain. The closer we can get to an olive being squeezed in a jar, the better it is for our brain and heart.

Even prioritizing healthy fats, we probably want to keep our intake moderate. Eating fat can increase our total blood cholesterol levels. Much like sugar, as we discussed in chapter seven, we do need some of it—just not too much and the right kinds.

Do You Love Guacamole?

Guacamole lovers rejoice. Avocados are a great example of healthy fat. Two avocados a week lower the risk of heart disease, a 2022 study found. It followed over a hundred thousand people for over thirty years and found that if avocados replaced animal fats such as cheese, butter, or bacon in the diet, there was a 16 to 22 percent lower risk of cardiovascular disease.[11]

What About Eggs?

When we hear "cholesterol," we often think of eggs. Eggs are one of the most dizzying debates in health. Eggs were a dietary villain, then they were a hero, and then a villain again. Where does this debate stand now?

As I mentioned in chapter three, there is good cholesterol and bad cholesterol. Your body needs some cholesterol to make hormones and vitamin D (more on that later) and to help you digest foods. In addition to eggs (specifically, their yolks), cholesterol is found in meats and cheeses.

The bottom line is that eggs can be part of a brain-healthy diet. Keep the eggs in moderation, which means about one or two a day for most people. Plus, once or twice a year, have a blood test to check cholesterol levels (as we covered in "Your One Sheet of Paper: Test Cholesterol" in chapter three). If the levels are too high, ask your physician about cutting back on eggs or just having the egg whites. If you currently have high cholesterol levels, just discuss egg intake with your cardiologist.

Healthy Proteins

If you've seen the amazing things that can be made from Legos, such as a re-creation of the Golden Gate Bridge or a Lamborghini, you probably already have a good idea about how to think about proteins. Proteins are the Lego pieces that, through different combinations and configurations, build you from head to toe. You use proteins to make hormones, key chemicals involved in learning and memory, and the antibodies of your immune system. Here's the deal, though: many of these proteins you can't store or make on your own. You need to ingest them. Luckily, they come from a wide variety of sources, including fish, eggs, soy, meat, milk, legumes, nuts, and beans.

One key type of protein is whey protein, which is found in dairy products. Studies have shown that taking whey protein as a supplement can help lower everything from LDL and total cholesterol to blood pressure.[12] Even people who are lactose intolerant have the option of consuming lactose-free dairy that contains whey protein.

By the way, many of the foods we think of as good protein sources are good for another reason: They also contain the nutrient choline. Choline is like a triple whammy of goodness to the brain and is critical for brain health and lowering the risk for dementia. First, it keeps the plaque-like trash in the brain from forming. Then, it calms down overactive microglia, keeping them from attacking healthy brain cells. Finally, the brain also uses choline to make acetylcholine, a critical neurotransmitter that allows the brain cells to communicate.

Although plant-based diets can have brain benefits, they come with a concern: vegan and vegetarian diets often lack choline. The food with the highest levels of choline is chicken livers. If you just gagged, don't worry; other foods loaded with this crucial nutrient include fish, chicken, pork, beef, eggs, shrimp, beans, milk, and broccoli. So, if you are living the veggie life, make sure to intentionally include vegetables that are high in choline.

You're Getting Sleepy . . .

We all like to sit around at Thanksgiving and talk about how the tryptophan in the turkey makes us feel sleepy. In reality, you would have to eat about thirty-five pounds of turkey to feel the sedating effects of tryptophan. (You might be saying, "Well, I do eat thirty-five pounds of turkey on Thanksgiving," but it only feels like you do.) What is actually making you sleepy are the fat- and carbohydrate-heavy foods like stuffing and mashed potatoes, overeating, alcohol, and, in some cases, the conversation.

Fruits, Vegetables, and Grains

Speaking of vegetables . . . let's talk about those next. It probably won't shock you to hear that fruits, vegetables, and grains are important parts of a brain-healthy diet. That's because they contain important vitamins, minerals, and fiber. You can add them to your plate in the following ways:

- Blend your fruits and vegetables for a powerful punch of nutrients.
- Eat vegetable-based soups. (Just beware of too much salt in canned soup—see page 187 for more on salt.)
- Toss vegetables into eggs and omelets.
- Add bananas and berries to oatmeal and cereal.
- Try veggie-based noodles.
- Throw some veggies on your pizza. And try a cauliflower crust.
- Pair a burger with a side salad or vegetables.

If you can look at your plate and see half covered with vegetables, you know you're on the right track. Now let's take a closer look at these plant foods and why they're so good for our brains.

Vitamin C

Vitamin C is involved in multiple aspects of immune health, including helping to produce white blood cells. These white blood cells protect against infection.

When we think of vitamin C, oranges often come to mind. But if I told you that one of these foods—broccoli, kiwi, bell peppers, or cantaloupe—has more vitamin C than an orange, which would you pick?

It was a trick question.

The answer is all of them have more vitamin C than oranges. (It looks like oranges have a great PR firm.) The point is that vitamin C is critical for immune health, and there are many foods besides oranges that provide a potent dose. The recommended daily amount is approximately 75 mg a day. For perspective, a single cup of strawberries has 95 mg of vitamin C. Interestingly, a study found consumption of four gold kiwifruit a day reduced the symptoms of a sore throat and head congestion caused by an upper respiratory tract infection.[13] The kiwis increased plasma vitamin C concentration in healthy older adults.

By the way, while prioritizing fruits and vegetables rich in vitamin C is a good move, there isn't much evidence that popping a vitamin C supplement is beneficial for general health, as the supplement can be difficult to absorb, and doses beyond 2,000 mg a day can upset the digestive system. (We'll discuss other supplements later in this chapter.)

Prebiotics (Fiber)

Prebiotics are food that the good bacteria in our guts (see chapter four) love to eat. (They're different from probiotics, which we'll address later in this chapter.) When you feed the good bacteria with prebiotics, the good bacteria replicate and grow. Some tasty prebiotics are tomatoes, artichokes, bananas, berries, flaxseed, legumes, chickpeas, walnuts, onions, garlic, chicory, dandelion greens, asparagus, leeks, and whole grains.

The key component of foods that feed good bacteria is fiber. Fiber has been shown to help with inflammation-related conditions such as heart health, blood pressure, and diabetes; it also decreases LDL (bad) cholesterol (see chapter three). Greater fiber intake is associated with more time in deep sleep.[14] What's more, prebiotic foods such as fruits, vegetables, and whole grains help release compounds into the blood that help you sleep.[15] Soluble fiber is found in oatmeal, rice, and oat bran, as well as citrus fruits, apples, pears, strawberries, peas, potatoes, and Brussels sprouts. (So, if you don't like Brussels sprouts, don't despair, you have plenty of other ways to get that fiber.)

We cannot digest fiber on our own, but when fiber reaches the gut, good bacteria feast on it. After they eat that fiber, good bacteria release butyrate. Butyrate heals and protects the lining of the gut.

If we don't have enough fiber to feed our good bacteria, bad bacteria take over and start munching on the gut's lining, causing inflammation. Only 5 percent of people in the United States meet the recommended daily target of 25 grams of fiber for women and 38 grams for men.[16] But adding fiber to your diet doesn't have to be complicated. Little changes can have a big impact.

- Instead of white bread, have wholegrain bread a couple of times a week.
- Swap brown rice for white, or try whole grains like barley, quinoa, or farro.
- Add legumes, beans, or nuts to meals.
- Toss a few extra fresh, frozen, or canned veggies into your soup or stew.
- Keep the skin on your potatoes.
- Throw some berries, nuts, or seeds onto your yogurt or salad.

A word of caution: eating too much fiber or adding a lot of fiber to your diet too quickly can cause gastrointestinal discomfort. Talk to your physician or a nutritionist to find the right balance of foods for you.

Quick Tip for When You Are Shopping

In the grocery store, just because a package says wheat or made with whole grains does not mean it has beneficial fiber. Look for food with labels that state 100 percent whole grain or whole wheat. Ingredients to look for are oats, oat bran, spelt, flax, rye, or barley flour. Look for the gold whole grain seal on food packaging.

Sulforaphane and Other Beneficial Compounds

Cruciferous vegetables (like arugula, bok choy, broccoli, Brussels sprouts, cabbage, cauliflower, collard greens, kale, and others) have a trio of compounds that work together to deliver benefits to your body and brain. Besides sulforaphane, which may help protect the brain, they also contain the compounds glucoraphanin and myrosinase. While these compounds may be hard to pronounce, they're easy to access: to have all three compounds activated and provide health benefits, you need to cut, chop, or chew the vegetable.

A study in the *Journal of Agricultural Food Chemistry* found that raw broccoli had ten times the amount of sulforaphane than cooked broccoli. Avoid boiling, roasting, or microwaving these types of vegetables. If you cook your vegetables, consider steaming them for one to three minutes, and keep the temperature below 284°F. Past this point, the sulforaphane can be destroyed.

But I promise, you can still cook your vegetables using two methods! If you want to roast your veggies, take a page from a study in *Molecular Nutrition Food Research:* researchers found that putting a little mustard seed or mustard powder, which is high in myrosinase, on these vegetables before roasting increases the powerful punch of the sulforaphane. With a myrosinase boost, even roasting doesn't kill off those beneficial compounds.[17] You can also try adding a sprinkle of mustard seed, daikon radish, wasabi, or horseradish to your cruciferous vegetables, as this sprinkle helps the chemical reaction take place and preserves the sulforaphane.

Another easy way to get a good dose of sulforaphane is to chop the vegetables and let them sit for about thirty to forty minutes before cooking. The chopping helps release the active form of sulforaphane and the half hour or so allows the key chemical to stabilize.

If you, like one of our former presidents, don't like broccoli, you might be tempted to try a broccoli powder supplement. There is concern that broccoli powder supplements often do not contain myrosinase, rendering the supplement ineffective. Try cabbage or Brussels sprouts instead.

Vegetables and NO Benefit?

Wouldn't it be awful to have to eat a vegetable you don't like and not even get the brain-boosting benefit? It can happen. Spinach is high in the antioxidant lutein, but it has the most protective effect if the spinach is *not* cooked. Cooking spinach destroys the brain-protective chemicals. Surprisingly, what's even better than eating raw spinach in a salad is *chopping* it or blending it in a smoothie because cutting the leaves releases more lutein.[18] For a bigger brain boost in your green smoothie, throw in a bit of healthy fat such as Greek yogurt to increase the absorption in the gut.

Sweet News for Chocolate Lovers

Dark chocolate with a cacao content of 70 percent or more has a high concentration of flavanols with anti-inflammatory properties.[19] Enjoy a small piece (about the size of a postage stamp) and know you're helping balance your immune system. Just watch the sugar content (see page 184).

Foods to Avoid and Limit

Now that we've covered the basics of what to eat, you're probably wondering what's off the table. I would never tell you to *never* eat something (variety is the spice of life!), but to protect our brains, there are certain things we should eat only in limited quantities.

What the Bad Bacteria Love to Eat: Additives, Preservatives, and Fast Food

Did you know there's an unwrapped Twinkie that's been on display at Chicago's Museum of Science and Technology since 2009? It still looks good enough to eat, and it probably is. But the additives and preservatives that keep foods like Twinkies from spoiling can wreak havoc on the gut (and the brain!) by increasing inflammation and confusing the microglia.

Fast food is often filled with additives and preservatives that feed harmful bacteria. A study published in the *American Journal of Clinical Nutrition* had a group of people eat burgers and fries—a typical fast-food meal. But half the group ate meals prepared with saturated fat, while the other half ate meals prepared with sunflower oil, a healthier fat. An hour after eating the meal, the researchers tested participants' concentration and focus. The group that ate the meal prepared with the healthy fat scored significantly better on attention tests.

A few weeks later, they brought all the participants back and switched the meal so that the people who'd had the high-saturated-fat meal now ate the meal prepared with sunflower oil, and vice versa. Again, the groups were tested, and those who ate the meal prepared with high saturated fat performed worse on the tests of focus and attention.[20] It is interesting to think about how quickly the food we eat can impact our brains. It's incredible that science created food that can't spoil—but these heavily processed foods belong right where that Twinkie is: behind glass in a museum, not in your gut.

Out of Sight, Out of Mind

Another good reason to keep processed foods in a museum? We are much less likely to indulge in unhealthy foods if they are not around us. It is challenging to resist that cookie if it is in the house. It's much less challenging to resist a cookie that you'd have to actively acquire. The trick is to try not to have these temptations in your kitchen. Try only to have these treats when it requires extra effort to obtain them, like when you go out to a restaurant or the ice cream store for a scoop. Even taking a walk is a great way to get that treat. And no, I don't mean the walk to the fridge.

Added Sugar

I've said before that sugar is not the enemy, *excess* sugar is. For instance, fruit has sugar. But when you eat a piece of fruit like a banana, you're getting not only a little natural sugar (fructose; about 14.4 grams of it in a medium banana) but also vitamin C, B6, and potassium, which are antioxidants needed for tissue repair, immune health, metabolism, and muscle function, as well as a healthy dose of fiber. The fiber slows the release of sugar into your bloodstream, so you stay full longer. A 2021 study found that eating two servings of fruit a day lowered the risk of developing type 2 diabetes by 36 percent, compared to those who consumed half a serving.[21] (This trend was found in whole fruit and not fruit juice. Fruit juice tends to have more sugar and less fiber.)

A can of soda, as a contrasting example, has a whopping 44 grams—or nine teaspoons—of sugar, usually injected in the form of high-fructose corn syrup.[22] It's actually amazing you can even dissolve that much sugar into a liquid. And by the way, you'd have to eat three bananas to get the same amount of sugar as you can find in one can of soda.

The American Heart Association says that men should have, at most, about seven to eight teaspoons of added sugar, and women should have about five to six teaspoons a day.[23] In the early 1800s, the average American consumed 45 grams, or about nine teaspoons of sugar, every *five* days. Today, the average American consumes twenty-two teaspoons of added sugar every day, which is almost four times the daily limit. That adds up to about 3,500 pounds of sugar throughout a lifetime. That's the equivalent of nearly two million Skittles. Even if they're your favorite candy, that's a lot of Skittles.

It's easy for those grams of sugar to add up, even if you think you're being careful—and usually, sugar will sneak in through packaged food. There was a time when fat was considered a "bad" food, and the food industry responded by flooding the market with fat-free foods. Now we know that healthy fats, such as olive oil and those contained in nuts, fish, and avocados, are brain-boosting and brain-protecting superfoods. Fat carries flavor, so if you take away fat, you have to replace it with something. The food industry chose to replace fat with sugar. For example:

- Fat-free chocolate milk has ten teaspoons of added sugar. (This one hurt.)
- Ninety-nine percent of flavored fat-free yogurt has 31 grams, or six teaspoons, of sugar.
- A low-fat ice cream cone can include six teaspoons of sugar.

In other words, even if you think you're picking the healthier item, double-check and make sure that sugar hasn't been snuck into it.

All this doesn't even account for the added sugars that hide under different names. Many products advertised as healthy, such as certain types of granola bars, yogurt with fruit syrup on the bottom, and certain seemingly nutritious cereals, contain high amounts of sugar. You can scan the ingredients list looking for *sugar* but come up short. Where is the sugar hidden? Look for:

- Dextrose
- Fructose
- Galactose
- Glucose
- Lactose
- Maltose
- Sucrose

Pretty much anything ending in *-ose* is sugar. Also, be wary of any product that includes syrup, such as:

- Agave nectar/syrup
- Brown rice syrup
- Carob syrup
- Corn syrup
- High-fructose corn syrup (HFCS)

HFCS is a big one, so let's pause here for a moment. If you see this ingredient listed on food packaging, drop the package and run. The food industry hired some mad scientists to develop high-fructose sugar in the 1960s.[24] This type of sugar was developed to be sweet, stable, and easy to use in packaged foods. It's an amazing scientific achievement that just so happens to be poison to our bodies and brains.[25] We now have data that HFCS disrupts the microbiome and increases inflammation.[26] A study at UC Davis found drinks with high-fructose corn syrup increased the risk of cardiovascular disease in young, healthy individuals in just two weeks.[27]

Let's get back to our list. In addition to *-ose* and syrup, be on the lookout for the word *juice*:

- Cane juice crystals
- Evaporated cane juice
- Fruit juice
- Fruit juice concentrate

And then there are the trickier ones like those below:

- Dextrin
- Diastatic malt
- Ethyl maltol
- Florida crystals
- Maltodextrin
- Barley malt
- Blackstrap molasses
- Caramel
- Honey
- Molasses
- Rice syrup
- Treacle

Who would have thought there would be so many names for sugar? Don't let the food industry pull this trick on you.

Excess Salt

Pass the salt or pass *on* the salt? We need to take many studies in the wellness world with a grain of salt, but the impact of salt on our health is not one of them. In one study, when people were fed a diet with the equivalent salt content of two fast-food meals a day, they had a more challenging time fighting off bacterial infection.[28] Dietary guidelines recommend limiting salt intake to about 2,300 milligrams a day, or about one teaspoon of salt a day, but much of the food we eat, especially packaged and fast food, has a lot of added salt.[29]

Salt, by any other name, is still salt: MSG, sodium citrate, sodium nitrate, and disodium phosphate are all salt in different forms. Try to avoid products with more than 200 milligrams of sodium per serving. Check the nutritional information when you buy packaged food. Another food industry trick: just because it says "reduced sodium" doesn't mean it isn't still far more than the daily recommended amount.

Foods to Avoid for Better Sleep

Don't make your brain prioritize digestion over sleep! Especially before bedtime, avoid heavily processed foods (foods with lots of additives and preservatives) and heavy foods (high in carbohydrates or starch). Some people may also want to avoid foods with a lot of spice. What is spicy can be based on the individual; one person can pop jalapeño peppers like they are Tic Tacs, and the next person can be in tears after a hint of pepperoni. If spicy foods are causing any indigestion, they can interfere with your sleep.

A study in the *Journal of Clinical Sleep Medicine* noted that a diet higher in saturated fat and added sugar was associated with less deep sleep and more waking throughout the night.[30]

Alcohol

The impact of alcohol on the brain and immune system is another area where the takeaways can be confusing and controversial. On the one hand, one glass of red wine a day is included in the MIND diet (discussed earlier in this chapter). There is evidence light alcohol consumption might lower the risk of Alzheimer's.[31] On the other hand, other research suggests any amount of alcohol might accelerate brain aging. For instance, a 2022 study found that just one drink a day made the brain look about six months older. Those who had four drinks a day had a brain that appeared about ten years older than their chronological age.[32] Do you have whiplash yet from the conflicting studies? Here's what we do know: heavy drinking or binge drinking can damage the brain and suppress the immune system.[33] (That's more than 14 drinks a week or 4 on any a day for men and more than 7 drinks a week or 3 on any day for women.) Moderate drinking—two alcoholic drinks a day for a man and one alcoholic beverage for a woman (based on average body weight for males and females)—doesn't seem to have much of an impact on the immune system, negative or positive. Some small studies show that a moderate amount of alcohol might help and balance the immune system.[34] It is possible that flavanols in red wine provide anti-inflammatory properties.[35]

When it comes to alcohol and health, it is also critical to consider any underlying conditions or genetic risk for Alzheimer's, as these can also compound or exacerbate negative impacts of alcohol. For example, any amount of alcohol may increase the risk of Alzheimer's in those who carry the ApoE4 gene. [36] One study found that light-to-moderate alcohol consumption was associated with learning and memory improvements in those that did not carry the ApoE4 gene. But those who did carry the gene and consumed light to moderate alcohol showed a decline in learning and memory.[37]

Here's the bottom line. Heavy drinking is not good for the brain. When it comes to moderate drinking, there are conflicting studies, but I would never say that someone needs to *start* drinking for their immune system or brain health. As it currently stands now, the impact of alcohol on the immune system is individual, based on genetics, underlying conditions, and environmental factors. The take-home message is avoid heavy drinking and discuss moderate alcohol intake with your doctor. Just don't have this conversation over a glass of wine, as that could be awkward.

What About Supplements?

There's a lot of conflicting information about supplements for a few key reasons:

- The FDA does not tightly regulate supplements, thus it can be challenging to know what is actually in the supplement and how much of the promised ingredient is present.
- Supplements can be classified as products that may support health as long as they have not been proven to cause damage. (That is a loophole that needs to be fixed.)
- Often the studies on supplements are funded by the company making the supplement or include just a small number of people.

Just because a study was done doesn't mean it's a good study. We always want to look for studies that are double-blind placebo controlled. *Double-blind placebo controlled* means both the subjects and those directly conducting the study do not know who is getting the placebo versus the treatment while they're collecting data—so there's no chance that any existing biases or expectations can influence the results. Unfortunately, many of the studies with supplements do not reach this threshold of reliability. The use of any supplements should be discussed with a personal physician. That said, there are a few supplements to consider for the health of your brain—and some you can skip.

Your One Sheet of Paper: Testing for Key Nutrients

When you're deciding whether to add a supplement to your daily regimen, get your numbers first! If your values are already in normal ranges, you might not benefit from adding those pills and capsules. At your next doctor's visit, ask about a blood test to check these key factors:

- Folate
- Vitamin D
- Vitamin B-12

Vitamin D

Unfortunately, over 40 percent of the US population has a vitamin D deficiency. Studies suggest vitamin D supplementation may lower the risk of developing colds and flus. When people with low vitamin D levels were given a vitamin D supplement, it reduced their risk of having a cold or flu by about 50 percent.[38] On the other hand, overdosing on vitamin D can have negative impacts, such as causing a buildup of too much calcium. Too much vitamin D and too much calcium can be toxic and lead to kidney problems and calcium stones. Thus, it is important to have levels checked with a blood test to determine if a supplement is needed.

Calcium

Calcium has multiple important roles in your brain and body. It regulates how you make neurotransmitters, and it's involved in memory formation. Calcium also directs blood flow to the brain by letting your heart know how much blood to deliver. The brain ages when there is too little or too much calcium in the brain and body.[41]

Sunlight: A Natural Source of Vitamin D

You may know that getting sunlight exposure naturally gives you a dose of vitamin D (specifically, when your skin gets sun, it makes vitamin D from cholesterol in the skin cells). A few things to keep in mind:

First, you need to be outside to get your vitamin D. The sun rays needed to make vitamin D are UVB rays, and windows filter out UVB. Getting natural light throughout the day from a window can be helpful for other reasons (such as helping you sleep in the evening, as we saw in chapter ten), but it's not a substitute for outside time to produce vitamin D.

Next, location and timing matter—as does skin tone. Studies suggest that the amount of sunlight needed varies at different latitudes due to sun intensity.[39] A study done from March to September in the UK found that nine minutes of midday sun exposure, three days a week, is sufficient for those with light skin. This was accounting for any weather during those months. An individual with darker skin could take thirty or forty minutes to produce the same amount of Vitamin D due to differences in absorption in darker skin pigments.[40]

Too much sun exposure can lead to premature aging and skin cancer, so protect your face and exposed body parts, including neck, legs, and forearms, with sunblock.

The recommended intake for calcium is 1,000 to 2,500 milligrams a day. It can be challenging to get enough calcium through food, especially if you don't or can't eat dairy products, which is why a calcium supplement may be necessary. And, of course, vitamin D is also used to help calcium absorption. But before you decide to add a calcium supplement, be aware that there is a concern, particularly, but not solely, for women, about too much calcium supplementation in conditions such as stroke or white matter lesions.[42]

One study followed dementia-free women aged seventy to ninety-two for five years. Scientists tested their memory, scanned their brains, and found that if they had white matter lesions (which are signs of vascular damage to the brain) and used calcium supplements, they were at an increased risk for dementia. If a woman with a history of stroke took calcium supplements, her risk of developing dementia increased seven times over not taking calcium supplements. For a woman with white matter lesions, her risk was three times greater if she took a calcium supplement than if she didn't. Women without a history of stroke or white matter lesions had *no* increased risk for dementia when taking calcium supplements.[43] Just remember that it is important to eat foods high in calcium, to be aware of your medical history, and to speak with your personal physician before adding a calcium supplement.

Fish Oil

We know that fatty fish like salmon is good for the brain. One of the areas that is murkier is fish oil supplements and their benefit on brain health. A 2022 study found that intake of fish oil supplements omega-3 PUFA (polyunsaturated fatty acids) was associated with lower risk of dementia among those between sixty and seventy-three years old.[44] Like with all supplements, ask your physician before adding fish oil to your regimen, but there is some evidence it might be beneficial in dementia prevention.

Turmeric/Curcumin

Turmeric is the spice that gives curry its bright yellow color. There is evidence that its active compound curcumin can have anti-inflammatory properties, and this might help brain health.[45] A double-blind placebo-controlled study published in the *American Journal of Geriatric Psychiatry* found that a curcumin supplement improved memory performance in people without dementia—those taking curcumin improved on memory tests by 28 percent over the eighteen-month period.[46] Furthermore, some of the participants taking curcumin saw an improvement in mood. The scientists performed

PET scans and found less tau and amyloid in those taking the curcumin. This was a small study of about forty people, and the results need to be verified in more extensive research, but at this point, there is enough evidence to say that turmeric/curcumin can be used as an anti-inflammatory, with some key caveats. It is essential to take the correct dose for the type of inflammation to be treated; for example, treating inflammation in the gut is a very different dose than inflammation in a joint, and dosage would need to be adjusted accordingly (a health professional can guide you). Finally, studies suggest that taking turmeric/curcumin with piperine, a pepper extract, increases absorption.

What's the Deal with Probiotics?

Probiotic supplements are pills or powders that contain known "good" bacteria (lactobacillus strains are particularly common). On the surface, it makes sense that supplementing good bacteria helps create a diverse microbiome, and a multimillion-dollar business has been built around that premise. But the scientific evidence that taking a probiotic supplement has any benefit to a healthy person is pretty much zero.[47] This is an area where marketing is ahead of the science, just like the pedometers and sleep apps.

Each person's microbiome is unique, and the truth is that taking a probiotic supplement when you have a healthy gut is like shooting a squirt gun into the ocean. The bacteria that already reside in your gut pretty much tell the new arrivals to keep moving along until they are eventually excreted. Furthermore, as noted at the beginning of this section, like other supplements, probiotics are not regulated by the FDA, and their makers can say their products are beneficial as long as they haven't been proven to do harm.

While probiotic supplements do not seem to benefit healthy individuals, there is evidence that they *can* benefit those with specific health conditions. For example, one meta-analysis found that probiotic supplements provided benefit to people with anxiety in 36 percent of cases.[48] However, the use of *pre*biotics (which we covered earlier in this chapter) provided benefit in treating anxiety 86 percent of the time.[49] As another example, consuming probiotics has been found to help people lose weight.[50] If you're interested in using probiotics for a health condition, speak to your doctor.

Unlike probiotic supplements, there is evidence that probiotics *found in food* can be of benefit.[51] Unpasteurized fermented foods such as yogurt, kimchi, sauerkraut, some pickled vegetables, and kefir contain probiotics. These types of food can populate the gut with good bacteria.

Not All Yogurt is Created Equal

When you're buying yogurt, opt for types without added sugar, which seems to feed the harmful bacteria. Yogurt with fruit already added to it often is loaded with a sugary syrup. Instead, add fruit like blueberries, raspberries, sliced strawberries, and diced peaches to plain yogurt when you're ready to eat.

The Big 5—and 3 Simple Questions

Diet is individualized and can quickly become overwhelming with the latest fad or trend. As complex and individual as the question of what a good diet is, at the end of the day, we can boil it down to a few basics: the "Big Five" foods to eat and three simple questions to keep you on the right track.

One way to keep it simple is to just double check you have these five items (or most of them) in your actual or virtual shopping cart. If you are consuming a combo of most of these five foods, you are likely going in a great direction:

- Fatty fish like salmon
- Avocadoes
- Nuts
- Blueberries
- Cruciferous veggies (don't forget the broccoli)

For an unexpected bonus, throw in some capers or red onions. Capers and red onions contain quercetin, an antioxidant that has been shown to have some heart- and thus brain-protective effects.[52] The amount of quercetin needed is still an emerging area of research, but some studies suggest two tablespoons of capers could give an added brain boost.[53] You could try throwing some capers on your salmon.

Beyond the Big Five, these three simple questions can help you determine if what you are about to eat is good for your brain:

1. **Will it spoil?** In many cases, perishable is a good thing! The additives and preservatives that keep food from spoiling wreak havoc on your gut bacteria and confuse your microglia. To protect your brain, think fresh.
2. **Are there tons of ingredients in that packaged food?** And for that matter, can you pronounce the ingredients on the package? Or does it look like the makings of a chemical experiment? OK, that's a few questions. But answering yes to any of these isn't a good thing. We do want to be careful about the pronunciation thing, though, because it can be tricky to pronounce some foods that are very good for you, like kimchi, kefir, or sauerkraut. And we want sulforaphane, which is a tongue twister. Think in terms of trying to avoid long lists of ingredients on packaged foods.

   Some examples of specific ingredients to minimize, especially if you have concerns about intestinal inflammation, include ethylenediaminetetraacetate (EDTA),[54] maltodextrin (MDX), and carboxymethyl cellulose (CMC).[55] Help your brain: eat minimally processed food as much as you can. And avoid anything where sugar is one of the first few ingredients.
3. **Do you see a rainbow on your plate?** Eat the rainbow! (Sorry, that doesn't mean Skittles. I know. I'm sad, too.) A mix of brightly colored fruits and vegetables provides the most benefit to the brain.[56] Different fruits and vegetables—prebiotics—feed different types of good bacteria, creating diversity in the gut. The brain-boosting

chemicals that give these fruits and vegetables their vibrant color have a synergistic impact when they are eaten together instead of being eaten alone.[57]

So don't think of feeding your brain as losing a food group. Think of it as gaining protein-rich fish and delicious, colorful fruits and vegetables. Eating to boost your brainpower shouldn't be dull, and your plate shouldn't be dull, either.

CHAPTER 15
Mind Your Environment

YOU ARE EXPOSED TO *10,000 TO 100,000* ENVIRONMENTAL CHEMI-cals and compounds in your lifetime depending on where you live, work, and play.[1] Not all of these are bad, but some could be aging and damaging your precious brain. Environmental pollutants are an underreported risk factor for diabetes, depression, and dementia. Here's the good news: A healthy immune system and liver can detoxify many of these chemicals. To put in perspective how critical it is to have a healthy liver, more than half of patients with liver disorders experience brain impairment.[2] A healthy liver is like an advanced cleaning system that keeps dangerous chemicals from entering your brain.

What's Your Exposome?

Your exposome is a concept from the field of genetics that describes all of the environmental factors, including chemicals and toxins, that you are exposed to in your daily life (such as through the air or water or in your food).[3] The exposome is still an evolving field, because trying to decipher the interaction and impact of hundreds of thousands of these chemicals is challenging and overwhelming. Still, while there is a lot about the impact of toxins and chemicals on our brain that we do not know, what we *do* know is that our health is affected by a complex mix of genetics and environmental factors.

We also know that childhood exposure to specific toxins has a profound impact on the health of the brain. Lead and mercury, for example, can build up in the body and damage the brain, leading to learning disabilities and behavioral issues.[4] Childhood exposure to these chemicals cost the United States *$7.5 trillion* in lost productivity and medical costs from 2001 to 2016.[5] While restrictions on the use of heavy metals have reduced childhood exposure to lead and mercury, during this time period, a million cases of intellectual disability in the United States could be traced back to exposure to things like flame retardants and pesticides.[6]

We want to protect our brains throughout our entire lifespan, and to do that, we need to be aware of the potential for exposure through the air we breathe, the foods we eat, the water we drink, and other materials in our environments.

The Air We Breathe

Did you know that when levels of certain air pollutants rise, the number of psychiatric admissions rises alongside? Toxins in the air can lead to psychiatric symptoms, including anxiety, and changes in mood, behavior, and cognition.[7] Air pollution can also reduce psychological well-being, which is essentially how you feel.[8] Point blank, what you are breathing in is impacting your brain age and the quality of your day-to-day life.

Most large cities deal with heavy traffic and the pollution it causes. One study analyzed 678,000 adults who lived in Vancouver and found a higher risk for developing dementia, Parkinson's, Alzheimer's, and multiple sclerosis when the adult lived less than 150 feet from a major road or less than 450 feet from a highway. The authors believe the increased risk is likely due to exposure to air pollution.[9]

A different study investigated older women who live in places with high levels of a specific type of pollutant called PM 2.5. PM 2.5 is thirty times smaller than one strand of human hair and is released from power plants or automobiles. When this type of pollutant was higher than standard levels, the women had an 81 percent greater risk of developing memory loss or global cognitive decline, and a 92 percent greater risk of eventually developing dementia or Alzheimer's.[10]

One particular study drove home the importance of air quality by looking at Mexico City—one of the most polluted cities in the world. A group of researchers analyzed whether this air pollution had an impact on the brain and found early stages of Alzheimer's and Parkinson's in the brainstems of Mexico City residents aged between eleven months and twenty-seven years old. This brain damage is startling, especially considering how young the subjects were. Researchers found these individuals' brainstems and gut nerve cells also contained high concentrations of toxic iron, aluminum- and titanium-rich nanoparticles—the same chemicals that spew out of the exhaust pipes of cars.[11] We know there is a pathway between the gut and the brainstem; this data suggests pollutants were breathed in and ended up in the gut. While not all substances can cross the blood-brain barrier, these pollutants can penetrate the brain: once in the intestine, they made their way into the bloodstream and entered the brain.

Toxins can also enter through our noses and have a direct ride to the brain. Once a toxin gets into the brain, microglia eat up this toxin to keep your brain clean.[12] But if the microglia are inundated and overwhelmed with too many toxins, they get confused—which, as we've seen, means they start to attack healthy brain cells and create inflammation. It's no wonder that levels of beta-amyloid (the stuff that can cause plaques) in the blood are higher in people who have been plagued by air pollution.[13]

The pollutants we breathe in also enter our bloodstream and disrupt the ability of insulin to manage blood sugar. Air pollution, which increases systemic inflammation, accounted for 3.2 million, or 14 percent, of all new diabetes cases globally.[14] And, of course, we know that diabetes is also a risk factor for dementia.

Cleaning our air will protect our brains. A recent study found that reduction of fine particulate matter (PM2.5) and traffic-related pollutants (NO2) led to lower dementia risk and slower cognitive decline in older women.[15]

A study in France found something similar: for every unit decrease in PM2.5, the risk of dementia dropped by 15 percent and of Alzheimer's disease by 17 percent.[16] Governments need to prioritize clean air as a crucial part of mental and physical health.

What Can I Do?

We may think that air pollution is a risk factor that's out of our control. Intriguingly, the study in Vancouver found that people could do something to lessen their risk significantly. Spending thirty minutes a day in a local park surrounded by grass and trees mitigated the increased risk of developing dementia associated with living near busy roads and highways. The authors speculate that the fresh air, natural air filters like grass and trees, sunlight, physical activity, and possible social interaction helped mitigate the risk.[17]

There are websites that can inform you of the air quality in your neighborhood on any given day. In some cases where air quality is very bad, it can be appropriate to wear a mask that filters out pollutants.

It's not just the air outdoors. Indoor air pollution is also a concern: household products and furniture can release toxins into the air. There are air quality meters that can be placed in your home. If the air is deemed not clean by the meter, try using an air purifier that contains a multilayer filter system composed of a prefilter, a carbon filter, an antibacterial filter, and a HEPA filter. Air purifiers can effectively remove some airborne pollutants and toxins in the house.[18] On days when outdoor air quality is good (and the weather permits), open your windows to air out your home.

Another factor that impacts indoor air quality is bedbugs. These common pests infest pillows, sheets, mattresses, and clutter. They release compounds into the air that negatively impact air quality and sleep (and, if you've ever had them, you know those bites itch). Besides removing bedroom clutter, frequent vacuuming, using a mattress cover, and washing bed linen in hot water, there is another often overlooked trick: put your pillows in the dryer set on high for about twenty minutes every few months. The heat kills the bedbugs and can help reduce the toxic compounds they release. These steps also bring the added benefit of clearing dust, which can also impact breathing and thus brain health.

As upsetting as it might feel to learn that you can't just hide away from pollution, do not be overwhelmed or stressed out. The simple actions we've just covered can lower your risk of toxin exposure.

The Foods We Eat and Water We Drink

Certain chemicals in our environment can alter gut health, leading to poor immune function.[19] These chemicals are everywhere—including in our food. So we need to revisit and expand our discussion of some of the foods we covered in the previous chapter. For instance, while fatty fish such as salmon can lower the risk of diabetes, if the fish is filled with toxins like mercury or pollutants, it will *raise* the risk of developing diabetes.[20]

Fish That Are High in Mercury:

- Shark
- Orange roughy
- Swordfish

Fish That Are Low in Mercury:

- Shrimp
- Canned light tuna*
- Salmon
- Pollock
- Catfish

* Albacore ("white") tuna has more mercury than canned light tuna.

That said, many of the simple steps you can take to lessen the impact of pollutants and counteract neurotoxins and inflammation involve food. The first food is fish—when it's missing the mercury. As we discussed in the previous chapter, fish is high in omega-3, which fights inflammation and neutralizes neurotoxins found in pollution. (What kind of fish was not effective? Fried fish. The frying process destroys the omega-3s. I am sad, too.)

A study investigated the amount of omega-3 in women's blood and correlated this to where they lived. If they lived in an area that was very polluted but they had higher omega-3, they had less brain shrinkage than women who lived in the same areas but did not have higher omega-3.[21] We have discussed that brain shrinkage is related to memory loss and dysfunction. Well, women who had higher levels of omega-3 had a larger hippocampus, the part of the brain involved in memory.[22] To get these benefits, you should eat more than one or two servings of baked or broiled fish or shellfish a week; that could counteract the effects of air pollution.

Besides fish, another food has been shown to lessen the impact of pollution. A study looked at nearly three hundred adults living in Jiangsu, one of China's most polluted regions. Half the participants drank a half cup of broccoli sprout beverage each day; the other half did not. Urine and blood samples were taken throughout the trial to measure the inhaled air pollutants. Those who drank the broccoli beverage showed rapid, sustained, and increased removal of benzene and acrolein, two toxic compounds that are common by-products of pollution and which increase cancer risk. The data showed that those who drank the broccoli increased the poisonous benzene's excretion and removal from the body by 61 percent and acrolein by 23 percent.[23] In short, the broccoli sprouts cleansed them.

Broccoli sprouts are wonder foods loaded with sulforaphane, which we talked about in the previous chapter. Studies suggest that sulforaphane activates a molecule in cells called NRF2, which helps cells remove environmental toxins. While sulforaphane is, again, found in cruciferous vegetables like broccoli, bok choy, and cabbage, and while both broccoli and broccoli sprouts are full of health-boosting compounds, there is evidence that the sprouts have fifty to one hundred times more sulforaphane than any other option.[24]

The cleanliness of water is also an important factor in brain health.[25] Many of the same contaminants just discussed in food can also impact water purity. Toxins and pesticides found in unclean water can have neurotoxic effects and increase inflammation.[26] Access to clean water can play a key role in overall and brain health.[27]

Organic vs. Non-Organic Food

Eating fruits and vegetables is a good choice, but be aware that large-scale industrial farming relies heavily on pesticides and fertilizers to maximize crop yield. Those pesticides and chemicals can remain on and in the foods you eat. Washing your fruits and vegetables removes dirt and surface pesticides, but there's the question of what remains in the food and how it affects both nutritional value and you.

Eating organic produce may reduce exposure to the toxic chemicals found in pesticides and fertilizers.[28] Organic can, however, be more expensive. To save some cost, consider the "peel rule": choose organic when the fruits and vegetables do not have a thick, inedible peel. So, yes to organic apples and strawberries and to non-organic bananas and oranges.

Organic food might also have more beneficial bacteria growing on and in it: An apple has about 100 million bacteria. Organic apples harbor a more diverse and balanced bacterial community, including abundant lactobacilli, one of the most known probiotic strains.[29]

You might have encountered someone who told you they could taste the difference between organic and non-organic. You might have thought this person was a food snob, but they might be on to something. Organic apples had significantly more of the bacteria called methylobacterium. This bacteria gives foods a pleasing strawberry flavor.[30] I can just hear Homer Simpson saying, "Mmmm. Methylobacterium. Delicious." So, an apple a day keeps the doctor away, especially if it contains good bacteria. That's one you don't usually hear, but a good one to remember!

And, by the way, eating other fruits and vegetables raw (when appropriate—raw potatoes or artichokes are neither tasty nor easily digested) can be just as good for you as the apple.

Notorious EDCs (and Other Materials in Our Environment)

So we've covered our exposures in our air and food, but there are more things to be aware of. If all of this is making you feel like you need to enclose yourself in a plastic bubble for safety, I have some bad news: that plastic itself might not be trustworthy. Why? Because of endocrine-disrupting chemicals (also known as EDCs). These chemicals, found in industrial products, plasticizers, and pesticides, can interfere with our hormones.[31] They can do serious damage, depending on the duration of exposure, a person's age at exposure, and extent of exposure.

One notorious EDC is bisphenol A (BPA), a chemical used in plastic packaging that can transfer to food. We don't really know how much of it is around us; one study found that the measurements used by the US Food and Drug Administration underestimate our exposure levels by as much as *forty-four times*.[32]

However, it is easy to tell if a plastic container contains BPA: just flip it over. Find the triangle recycling symbol, three arrows around a number between 1 and 7. Items with numbers 3, 6, and especially 7 are most likely to contain BPA. Items with 1, 2, 4, or 5 generally do not contain BPA.[33] Try not to use plastic that has BPA in it, and, no matter what number you see on the bottom, beware of leaching. Leaching occurs when canned or plastic foods are stored at high temperatures, like those in an un-air-conditioned car on a hot day. The heat can cause the chemicals in the plastic to leach into the food and then enter your body. This is why you should avoid microwaving food in *any* plastic containers.

Parabens are another EDC. These chemicals are used as preservatives and are found in many commercial foods and cosmetic products. Look for methylparaben, propylparaben, and butylparaben in the ingredients list on the back of the bottle or container. Those are three of the most common paraben ingredients to avoid.[34] A concern is also not just the amount of parabens in one product but the cumulative effect of multiple products containing these chemicals.[35]

Cleaning and Household Products

An interesting study had housekeepers wear backpacks that measured exposure when cleaning with common household products versus green products. The switch to green products meant a significant reduction in seventeen cancer-causing and hormone-disrupting chemicals, including benzene and chloroform—there was an 86 percent decrease in chloroform exposure with green products.[36]

There is emerging evidence that certain household cleaners, detergents, and personal care products that contain phthalates and polyfluoroalkyl substances (PFASs) can disrupt gut bacteria balance in kids.[37] This also raises concerns over how these products impact the brain and overall health throughout one's lifespan. Try to choose products that are "eco-friendly" or "green"—which may be harder than it sounds, unfortunately. It's not easy to identify which household products are actually safe for you and the environment, because even products labelled green don't have to adhere to strict standardized rules. One thing you can do, though, is to look for the Safer Choice label. For some cleaning jobs around the home, try a safe combination of water, vinegar, and dish soap instead of using any cleaning products at all. Other ways to reduce exposure and protect yourself from cleaning products is to open windows and doors to increase airflow while you're cleaning, and to wear gloves and goggles.[38]

The key take-home message is that although there are aspects of our environment we cannot control, there are actionable steps you can take to

Can You Sweat Out Toxins?

A booming industry promises to sweat out toxins through exercise, hot yoga, and saunas, but a study found the amounts of pollutants excreted in sweat to be small.[39] To put this in perspective, even with forty-five minutes of high-intensity exercise, you'd only sweat 0.02 percent of what you ate and drank that day. Although exercise is terrific for many reasons, and saunas may have other benefits like lowering blood pressure and stroke risk, don't turn to them for detox. Sweating is ineffective in removing pollutants from the body.[40]

leverage what you *can* control to protect your brain. The goal of this chapter was not to cause panic or overwhelm you with a long to-do list. Instead, just keep a few things in mind—they'll go a long way:

1. Be aware of the air quality in your environment. Use an air purifier when needed, and give yourself time each week in unpolluted low traffic areas, like a park.
2. For fruits and vegetables that don't have a thick peel, try to buy organic.
3. Try to choose home and self-cleaning products that are not loaded with chemicals. There are lots of options for these products now. Remember, clean air and clean food can lead to a clean brain.

CHAPTER 16

Cross-Train Your Brain

YOUR BRAIN CAN REMEMBER ABOUT 2.5 PETABYTES OF DATA—which is about the amount of data that could store 300 million hours of television. That's the equivalent of watching TV twenty-four hours a day for thirty-four thousand years. That is a lot of binge watching! Just imagine how far you can go to age-proof your brain by filling those petabytes with new information—whether you're binge-watching documentaries, listening to podcasts, taking up a new skill . . . or reading a book like this one. In fact, there is a lot you can do to get the most out of your brain.

Hopefully you've spent this entire book learning new information. That very action was protecting your brain: learning new information plays a significant part in disposing of brain trash via a "power wash" that uses one of your body's most effective brain cleansers: norepinephrine.[1]

Norepinephrine is a hormone and neurotransmitter that regulates heart rate, attention, memory, and cognition.[2] When you learn something new, your brain squirts norepinephrine from a brain structure called the locus coeruleus.[3] The norepinephrine breaks up the waste and trash in your brain so it can be excreted when you sleep. This keeps your brain young, healthy, and able to make new connections.

Maintaining a healthy brain and strong memory is not all about Sudoku, crossword puzzles, and brain games. Those can be fun, and they can have a brain benefit, but they aren't the only things we can and should do to exercise the brain. What does make significantly new connections (synapses) in our brain is learning new skills and acquiring information; the more connections you make, the more likely you are to retain and even enhance your memory. Your brain needs—it *craves*—more when it comes to stimulation.

Study after study has found that learning new things significantly lessens the risk of memory loss and slows down brain aging. For example, one study found that more years of education were associated with more active frontal lobes when adults took memory tests.[4] Activity in the frontal lobe is associated with better memory. But higher education isn't the only way to maintain memory; what we do day to day makes a difference. Another study found that even if individuals had lower levels of education, if they attended lectures, read, wrote, and, yes, played word games or puzzles, they had memory scores on par with those with more education.[5]

Learning a new language or a musical instrument is good for your brain. A study found that being bilingual or a musician at any age made the brain more efficient in day-to-day memory tasks.[6] But to get a real good spray of norepinephrine, make the new information challenging. Embrace that feeling of frustration when you learn something new and outside your field of expertise.

When you think about learning something new, approach it the way you would fitness training. For instance, you wouldn't go to the gym and only work out your forearms. (Eventually, you would look like Popeye.) You want to work out different muscles on different days and work on your aerobic fitness as well as building muscle.

The same goes for the brain. Learning a language works out different parts of the brain than sports and music do. You can cross-train your brain by mixing mental and physical learning activities: over the course of a week, for example, play tennis, golf, pickleball, or soccer; learn a new song to sing or play; and try your hand at a new language or read a book on a subject you're not familiar with.

Learning new things is one of the best things you can do for your brain; it also helps improve focus, productivity, and even creative thinking.[7] The rest of this chapter is dedicated to breakthrough tips to get the most out of your brain and the new things you are learning—practical steps to boost your focus, remember more, and maximize your creativity, all based on brain science.

Get a Focus Boost

Have you ever had this experience? You open a book and start reading. When you look up, two hours have gone by. You've been in a flow state, the term for those moments when we are so focused that we completely lose track of time. Deep states of focus are critical to boosting productivity and remembering more of what you learn. But how do you get there?

A 2017 study followed a group of people taking a test. Half the participants were asked to place their phones on their desks as they took the test. The phones were turned off. The other half had been asked to leave their turned-off phones in another room. Those with their phones in the different room scored on average 20 percent higher than the other group—our phones can be *that* distracting.[8] Just seeing your phone can be enough to make your brain wonder what you're missing—did you just get a text message, a post, or an email? This kind of micro distraction can have a significant impact on brain performance. Does reading this paragraph make you want to check your phone?

Your brain is wired to pay attention to new information because, for millennia, it was essential for your survival. Now, thousands of years later, every time your phone or computer dings with a new message or email, your brain squirts out dopamine. We love that feeling, but it can lead to mental exhaustion. These devices are amazing, wonderful, and powerful, but can be a distraction, and distraction is the enemy of productivity because our brains love novel information. The information can be important and meaningful or it can be a waste of time. Our phones beep, buzz, and provide unexpected information . . . just like a slot machine, keeping us hooked for the next dopamine hit. Slot machines steal our money, and our devices can steal our focus and precious time. And just like when we're sitting at a slot machine, it can be hard to realize the hours spent as the wheels spin and not much is getting accomplished.

The good news is that you don't have to be controlled by your devices. You can get your focus back by using your knowledge of how your brain works, and how you do it is surprisingly simple: step away from your phone, your computer, your tablet. If you want to get in the deepest levels of focus, place your phone (or computer or tablet) out of sight, so it is out of mind. And if your work needs to be done on the computer, try disabling your internet connection or only having one working window open. In order for me to write this book, I had to practice these same techniques. I put my phone in the other room and disconnected my WiFi while I was writing on the computer. Then I would set a timer, as I describe in the next section. Remember, brain scientists designed cell phones, websites, and apps to hijack the dopamine reward system so we wouldn't put our devices down, but we can outsmart these sneaky scientists.[9]

Mental Gymnastics to Exercise the PFC

Now that you have removed a key distraction, it's time to take your brain to the gym and exercise your focus muscle, your prefrontal cortex. You can do that through mindfulness, as we mentioned in chapter eleven. But there is also a focus and time management technique developed in the late 1980s called the Pomodoro method, which is one way of working out your prefrontal cortex. First, eliminate all distractions: silence your cell phone and put it out of sight. Silence the alerts on your computer. No websites open in the background. No checking email, the score, or social media. Set a timer (not on your cell phone!) for twenty minutes and focus on just one important task. If your mind wanders, bring it back to your task. Commit to twenty minutes of distraction-free work.

If you make it, be happy for yourself and consider that one rep. Take a five-minute break and do it again. If you are having trouble reaching the twenty minutes without a distraction, try just ten minutes, and each day you are successful, try increasing the amount of time until you hit twenty minutes of pure focus. You can definitely go past twenty minutes if you find yourself in the zone, but twenty minutes is the minimum. Again, think of this as if you were in the gym: increase the time and number of reps—periods of targeted focus—and get stronger and grow your prefrontal cortex. There is no magic formula for the number of reps you need to do, as it is based on how much work you need to get done that day. You can determine how many reps you need to do a day.

Don't forget that five-minute break, though. A 2011 study looked at the benefits of distraction. Participants in the study were separated into two groups. One group focused on a task for fifty minutes with no breaks; the other group took two breaks, one every twenty minutes. The group that took the two breaks within the fifty minutes remembered more about the task.[10] It seems that your brain needs a little distraction to reengage focus; it can be as simple as standing up and walking around the room or a quick stretch.

Remember More

Now that we've honed our focus to take in information, how do we remember more of it? As I mentioned earlier, practice makes (almost) perfect—and a lot of it has to do with repetition. But there are some other things you can do to help the new brain connections stick.

Say It Out Loud

We covered this in chapter five, but when you want to remember something, say it out loud. The next time you park your car and want to remember where it is, stop for a moment and say out loud something that identifies the location.

Remember Like Billy Joel: Make an Association

Daniel Tammet is a savant who has been studied by brain scientists all over the world. He memorized and recited 22,514 decimal points of the mathematical constant pi and learned ten languages in a week.[11] As a savant, Daniel has incredible memory, but he struggles with other aspects of basic memory. He also has synesthesia, a neurological condition in which information that stimulates one of your senses stimulates several senses. When an individual with synesthesia thinks of a number or a letter, for example, their brain automatically gives it a specific color, sound, or taste. Letters or numbers may embody distinct personalities and genders.[12] This condition is linked to increased creativity and is found in artists and musicians.

For example, when Billy Joel hears music, he sees colorful lights associated with each note. Individuals with incredible memory often either have synesthesia or use techniques that mimic the condition. Why? Because they are associating what they are learning with senses, feelings, and emotion. Remember, each of the five senses is stored in distinct regions of the brain, so using all of them makes the memory more likely to stick.

If I ask you to remember a family member's home, it might remind you of a specific smell, such as a meal cooking in the kitchen. For example, the memory of my grandparents' home always reminds me of a distinct type of soap. Just smelling this soap brings back a flood of memories. Tap into your powerful sense memory and associate what you are trying to remember with a sense or feeling.

For example, if someone tells you their name, associate their name with someone famous or a place you have an emotional connection to. Let's say someone introduces themselves as Henry. Imagine King Henry and picture the person you just met with a crown on their head. Take it a step further and place Henry in an English pub. Conjure up what the pub smells like. Is it cold in the pub or warm? These techniques might sound silly, but the more you associate the new information with past knowledge and build a story that taps into your senses, the more likely you are to remember it. These same techniques of associating and tapping into sense memory are used by memory champions to remember astounding amounts of information.

Make It Emotional

Have you ever found that you can't remember something important you need to do today but you easily remember an embarrassing moment from a bad date years ago, down to the exact words you said? Am I making you cringe? Sorry about that. We tend to remember things that are tied to emotion. Our brains evolved this way for our survival.

One of your brain's most important jobs is to keep you from experiencing pain. To do so, you must remember the things that bring you joy or cause hurt. In the diagram below, find the amygdala, the emotional part of your brain. Notice the amygdala is right next to your hippocampus.

Remember your hippocampus? Information goes there first, and your brain decides if it's important enough to remember. If it is, the memory is sent to different parts of the brain, depending upon what type of memory it is. Take passwords, for example. Passwords are stored way out at the exterior of your brain. This is a long way for the information to travel. (So, if you're getting upset that you keep finding yourself hitting the password reset button, don't feel bad. Your brain wasn't designed to remember random words, letters, and numbers.) But if it's an emotional memory—something that

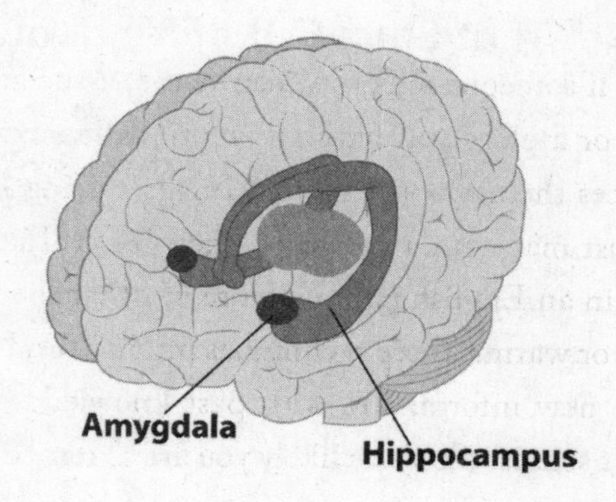

makes you feel—it goes right next door to the amygdala after leaving the hippocampus. The less distance the information must travel, the less likely it will get lost along the way. It is also important to note that the hippocampus is involved in recalling memories and also helps coordinate finding information. If you associate the information you are trying to learn with something emotional, it will help you store and recall the information.

If you want to remember a password, make it emotional. Use a word or phrase that means something to you. Something silly. Something scary. Something sentimental.

The emotional connection is also (part of) why music is so powerful for memory. (For instance, we never forget that "the best part of waking up" is a certain brand of coffee in our cup.) Songs and jingles have real sticking power in your memory because music is stored all throughout your brain.[13] Music has emotion, which is stored in the amygdala. Music also has rhythm, which is stored in the motor part of your brain (more on that shortly). And music—obviously—has sound, which is stored in the auditory part of your brain. Again, the more places a memory is fragmented and then stored, the more likely you are to retrieve the memory and remember (and maybe if I repeat that enough, you'll store that information in multiple places, too). If you are trying to remember something, make it musical and give it a melody.

Get That Song Out of My Head

We've talked a lot about remembering, but what about something you want to forget? Have you ever had a song on repeat in your head? These are called involuntary musical memories or earworms. When you get one, maybe you're happy because it's a song you love, or maybe you've heard "Who Let the Dogs Out" enough for one day. A study found a trick to get that song out of your head: Chew gum.[14] The action interferes with this involuntary memory. There's something to chew on.

Here's another tip: If you have a piece of information or fact that you really want to stick in your memory, make up a silly rhyme either about or that includes the information. Yes, it might sound absurd, but you are engaging the emotional part of your brain, and that information will stick. Another important tip, if you use the silly rhyme technique to remember someone's name, don't say it out loud to the person you just met. It can get very awkward, very fast.

Tap into Motor and Visual Memory (or, Why You Never Forget How to Ride a Bike)

Motor memory is the memory of movement.[15] A form of this type of memory, non-declarative memory, is exceptionally strong, and it is the reason why we never forget how to ride a bicycle. It's also why, in most cases, people who have amnesia don't forget how to button their shirts or tie their shoelaces; motor memory often is not damaged in amnesia and some cases of dementia.[16] There are several examples of musicians who developed amnesia, including Clive Wearing, a famous piano player. Clive could still remember how to play all of the songs he knew before he developed amnesia, but he couldn't tell you the names of any of the songs or where he learned them.[17] This is because playing the piano involves motor memory, which uses different areas of the brain from those involved in remembering information like facts or names of songs.

Furthermore, if people try to memorize a list of words, they remember more if they draw a picture that relates to each word.[18] Those little doodles employ several critical factors that help us remember:

- Drawing is a *motor* skill.
- It takes *time* to draw, which helps your hippocampus realize that the information is important.
- Drawing a doodle helps us *visualize* as we conjure up what the item we are drawing looks like.
- When we draw, we often tie an *emotion* to the drawing.

Studies with younger and older adults found improved memory when participants made a little drawing of words they were trying to remember, because they were tapping into motor memory.[19]

Just taking a moment to visualize any information you are learning is a powerful technique. Close your eyes and build a mental picture. The visual part of the brain is exceptionally strong. To revisit some notable people we met early in this book, researchers have found that SuperAgers highly activate the visual part of the brain when remembering and recalling information. It is believed this is one of their memory secrets.[20]

Often people will say, "But I just want to memorize the information. Do I really need to do all of these tricks?" That is much like saying, "I want to look like Arnold Schwarzenegger in his prime, but I don't want to go the gym." Now, we don't need work that hard. But for some information, you do need to put in a little effort to let your hippocampus know that the

Mnemonics

A mnemonic is a memory technique to help your brain remember information. Here are a few you may have seen:

ROY G. BIV: the colors of the rainbow—red, orange, yellow, green, blue, indigo, violet.

HOMES: the names of the Great Lakes—Huron, Ontario, Michigan, Erie, and Superior. *Super Man Helps Every One*: the names of the Great Lakes from biggest to smallest and west to east.

To improve your memory, I came up with a mnemonic based on all the memory techniques we just covered. Remember them with SAVED:

Say it out loud.
Associate something with it.
Visualize it.
Emotionally connect with it.
Draw it.

information needs to stick. Have fun with these games and try not to giggle as you find yourself imagining the person you just met riding an elephant down the Las Vegas Strip.

Maximize Creative and Innovative Thinking

I found a video online of two of my favorite musicians talking about how a great song is born. Billy Joel asked Don Henley, "When does the song come to you? Is it when you are sitting at the piano or the guitar, laboring for hours on a sequence of notes or a melody?" Henley replied that his best songs rarely happen that way. Instead, the spark of inspiration sometimes arrives when he is washing dishes. Joel agreed that his "a-ha moments" almost always arise when he is washing dishes or partaking in mindless repetitive activities as well. Henley added that he sometimes washes dishes in hopes that a great song idea will appear. At the very least, he said, his wife is happy that he did the dishes.

As I watched their discussion, it reminded me of a series of studies on how the brain innovates. One such study, by neuroscientist and psychologist Nancy Andreasen, featured fascinating research on thirteen groundbreaking, game-changing individuals in a variety of different disciplines and fields. We are talking Nobel, Oscar, and Pulitzer Prize winners—filmmaker George Lucas, writer Jane Smiley, mathematician William Thurston, and six Nobel laureates in the sciences. She wanted to find out whether all these individuals had the same process to get a great idea.

Andreasen and her team followed the subjects around, scanned their brains, and conducted extensive interviews. In some cases, they even moved in with the subjects to track their every move. Andreasen and her team hypothesized that the innovative thinking process would be different based on the subject's occupation; for example, someone in math would have a different process than a writer or artist. Surprisingly, after studying all these individuals, *it essentially didn't matter what their field was*—they all followed a very similar and deliberate formula that led them to their breakthrough moment.[21]

The Secret Formula and the Power of Downtime

Interestingly, the Henley-Joel interview shed light on an underlying problem that's unique to our time: even while I was trying to focus on the discussion I was watching, YouTube displayed other videos I might be interested in. At the same time, my cell phone was buzzing and beeping with texts, emails, and notifications. And I kept being reminded of the other things waiting for me on my to-do list.

Now, if I asked you to picture a high-producing employee, would you picture someone who comes in before everyone else and doesn't leave until the cleaning staff arrives? Someone who, during those hours, is talking on the phone and typing away on the computer (with multiple apps and websites open at the same time) while simultaneously brainstorming with a colleague in their office? This multitasking, overtime-working individual is often the image most of us have when we think of someone who is productive. They aren't waiting for inspiration or the a-ha moment to arrive; they're actively pushing their brain to its maximum, chasing down a breakthrough. This seems to make sense, and if ideas were like fishing, the big fish would come to those who spend all day on the water, dropping in multiple lines and tirelessly tending to each pole, right?

Notice the word *seems.* The truth is that innovative ideas come when you are *not* pushing your brain to the maximum all day long.

These days, we are all immersed in an endless stream of information. Not just a twenty-four-hour news cycle and the availability of television and streaming media at any time, but constant texts, emails, posts, tweets, pings, and alerts.

Rather than help us get more done, this merry-go-round of information spins around us as we also try to juggle work tasks. A typical modern workday feels like playing Whac-A-Mole as our attention jumps from one task to the next. Swimming in this constant information stream, we can quickly feel exhausted and unproductive, even if we did accomplish a lot. Experiencing this feeling day in and day out is a fast track to burnout. This is a serious issue: approximately half of all employees say they are experiencing burnout.[22] Those with higher levels of burnout have a 79 percent increased risk of heart disease.[23] We have discussed in-depth the link between heart disease and brain health.

Our brains are good at convincing us we are doing great even when our brains are struggling to make sense of too much information. There was an experiment in which subjects sat in a car simulator and "drove" while dealing with distractions like text messages.[24] Most subjects reported that they aced the drive with flying colors. They're wrong. Most people ran through stop signs and swerved into other lanes; in real life, they would have risked arrest or caused a serious accident.

When it comes to learning new things, being productive, and improving innovative thinking, we can break down maximizing our brains into two key steps:

1. Optimize your brain's ability to take in the most information in the least amount of time.
2. Take new and previously stored information and use it to create and innovate.

A crucial part of the process to maximize your brain is to have a healthy brain, of course, but just having a healthy brain alone is not enough for your brain to realize its full potential. Let's go back to the Andreasen study. Andreasen found that each of the subjects spent some time each day in deep states of focus, *only focusing on one task that they wanted to accomplish in their respective field*. No multitasking.[25] Distraction just doesn't give us the mental space for deep or creative thinking. We need to take one thing at a time.

Inspiration doesn't arise out of nowhere; it's rooted in years of focus, what the author Cal Newport terms "deep work." Sir Isaac Newton wasn't the first person to be hit in the head with an apple: it was the deep work he'd done in the decades leading up to that moment that left him with more than just a bruise. Billy Joel and Don Henley spend hours grinding away at the piano and guitar, playing and writing, in uninterrupted deep focus. And then, when they least expect it—at the dishwasher, in the shower, driving to the store—inspiration strikes, and they create the iconic songs we have come to love.

The formula that Andreasen and others uncovered is that to maximize productivity, the brain needs to balance spending time focusing on a single task *with stepping away and relaxing.*[26] It's hard work *plus* a break that enables our brains to process and play with information that optimizes our brains.

In some ways, your brain is like a computer. As a laptop can update an app or scan for malware while you're working, your brain can work on problems while you're not thinking about them. Have you ever struggled to remember the author of that great book you just read, or the actor in the movie you just watched? The name feels like it's on the tip of your tongue or floating just beyond your grasp, but it eludes you. Then hours later, maybe when you're walking down the street and not thinking about that book or movie at all, the name suddenly occurs to you. That is an example of your brain solving problems for you in the background.

Your brain is powerful beyond your conscious awareness. You know things, and you don't even know how you know them. But to access that power, you need to be able to relax. This isn't so simple in a society where we are rewarded for looking busy—for running on a hamster wheel instead of getting the best out of our brains. You can't just work hard. You *must* also relax wholly and completely. That formula not only helps you learn and create better but also helps you keep your brain healthy.

So, how do we give our brains a real rest? After a session of total, singular focus, do something you enjoy, and *only* do that one thing. Also, turn your phone off. Better yet, turn it off and leave it in a drawer. Now that you have ditched your phone, do something completely different:

- Go for a ten-minute walk.
- Do a quick workout.
- Play a sport (if you're a professional or competitive athlete, choose a sport you don't usually play).
- Get a massage.

- Vacuum or wash dishes (preferably at my house).
- Swim.
- Listen to music.
- Play a musical instrument.
- Do a breathing exercise.
- Or simply close your eyes for a few minutes and do absolutely nothing.

If you are like me, you might have found yourself with a lot of windows open on your computer or phone. If it wasn't working like it used to, you probably called IT or tech support. Did you notice they pretty much have one piece of advice? Turn off your device. Wait a few minutes and turn it back on again. That is great advice for the brain as well. There comes a point in the day when we just need to power down and take a break to reach the full potential of the brain.

Your True Potential

Mastering the steps to optimize your brain and creativity takes time. The steps may seem simple, but implementing them is not always an intuitive process. Carve out a little bit of time each day to:

1. Decide what information is of value to you. Everything else is a distraction, so consciously eliminate as much diversion as possible.
2. Set a timer and commit to at least ten minutes a day of deep focus on one important task. Once you have mastered ten minutes, try to increase to fifteen minutes and then twenty to twenty-five minutes. (If you can jump straight into the Pomodoro method, go for it!)
3. Give your brain the gift of total relaxation for a few minutes after you complete your task.
4. Repeat.

Keep in mind that step 3 probably won't deliver a full, fleshed-out solution to whatever it is you're struggling to discover. That's the time when the brain delivers part of an idea or innovation in a quick burst; you can take that idea back to deep, uninterrupted focus and tinker with it. Most breakthrough ideas take many, many rounds of this cycle to be fully realized.

There is one more piece to this puzzle. Nancy Andreasen discovered in her study of highly successful, creative people that besides the formula just outlined, there was another commonality: perseverance. Her subjects did not quit.[27] This is also true with keeping your brain youthful. It's all about little steps each day and not giving up.

With this simple practice, you might likely be surprised how your brain will soon deliver innovative ideas that excite you beyond anything you've experienced. These "a-ha moments" are your reward for treating your brain how it deserves to be treated. I am excited to think of all the wonderful ideas your brain has in store for you. And, with the strategies you've learned in this book, they will keep on coming for many years into the future—all thanks to your balanced, healthy, age-proof brain.

CONCLUSION
A Message of Hope

WE STRIVE FOR SO MANY THINGS IN LIFE, SUCH AS HEALTHY RELAtionships, happiness, a strong body, a successful career, a good buffet, and a gripping show to binge-watch. But if we don't have our brain health, we don't have much. Fortunately, as more and more studies from across the globe confirm every day, we are not held hostage by our genetics, and we have power over our brains and immune health.[1]

We do not—yet—have a cure for dementia, Alzheimer's, depression, or anxiety. Still, we have strong evidence that simple lifestyle interventions can dramatically improve brain health and lower the risk of disease today, tomorrow, and in years to come. In Finland, a thousand people aged sixty to seventy-seven who were at high risk for Alzheimer's were put on a program that included the elements of brain health I cover in this book—changes to their diet, sleep, and exercise, as well as managing underlying conditions. Their scores on cognitive tests increased 25 percent, their ability to manage their days went up 83 percent, and their brain processing speed more than doubled.[2] Studies at Weill Cornell Medicine found that when individuals followed protocols covered in this book, they showed signs of reversing early stages of memory loss.[3]

The world of wellness is filled with bogus solutions and empty promises. We are living in a time when there has never been so much information—or misinformation. Among the overwhelming noise and mess, this book has given you simple, scientifically valid, concrete, and actionable steps that can significantly protect your brain:

1. Prioritize sleep to wash your brain.
2. Learn new things to break up the brain trash and make new connections between brain cells.
3. Be socially engaged and treat hearing loss.
4. Embrace acute stress and use mindfulness to manage chronic stress.
5. Put out the fire of inflammation by managing your diet and stress.
6. Eat food that spoils (just not when you are eating it). Minimize packaged food with ingredients that look like they are for an advanced chemistry experiment.
7. Make moderate exercise, such as thirty minutes of walking, a daily habit.
8. Treat diabetes by consulting a physician.
9. Minimize contact with toxins by using green cleaning products and giving yourself some green time with nature.
10. Take care of your heart by being on top of your blood pressure and cholesterol.
11. Prioritize and treat any mental health issues.
12. Utilize the "One Sheet of Paper." Pass some of this responsibility to your physician. Tell them you want to be on top of your brain health, now and years down the road.

While you were reading this book, you did at least one of two things on this list: either you learned something new, or this book put you to sleep. Either way, it was good for your brain!

All jokes aside, I hope you close this book feeling empowered. One of my biggest messages is that healthy brain habits that impact long- and short-term brain health are made through small steps each day.

Trying to commit up front to twelve healthy habits for weeks and months can be a recipe for quitting. Some people look at this list and say, "Yes, I have twelve things! Let's do this!" Others say, "That feels like a lot. Where do I begin? Maybe I'll just take a nap first." (If you take a nap, remember to make it 30 minutes or less!) Instead, make the goal to commit to any three of these healthy habits each day (there are more details and suggestions in the appendix, too). Wake up each day and focus on what you can do that day. For example, go for that morning walk, throw some veggies on your plate, and floss your teeth. If that's too much, start with one of these brain-boosting changes. Once that change becomes a habit, add another and then another.

We have discussed many studies, but I want to leave you with one more; it's one of my favorites. Scientists wondered if there was anything that was better than walking for making the brain look younger. The researchers took a group of participants in their eighties and found one activity that beat it: dancing.[4] Whether it's ballet, ballroom, or hip hop, and whether you're "good" at it or not, learning to dance involves learning on multiple levels. It is physical exercise. It's coordination. It engages your mental resources and memory. It can be social. It involves hearing. It's stress relief (unless, perhaps, you're dancing with me). If you are thinking, "No! Not dancing. *Anything* but dancing," have no fear—it doesn't matter if you are a bad dancer. If you really don't want to dance, yoga or tai chi, which also incorporate multiple levels of brain-protective activity, work, too.[5] And studies have found that playing sports and singing also hit many brain-boosting factors on our list. (But 'fess up: Don't you secretly want to do a little dancing? How about when no one is looking?)

This brings me to a last key message. The topics of brain health and dementia are extremely serious and intense and can be understandably stressful and anxiety provoking. But in all the seriousness and complexity of brain science, there's a simple truth. The more we understand the amazing brain, we realize how important it is to include and even schedule fun into your daily life. Sing, dance, learn new things, be social, play a sport, and embrace hobbies and loved ones. Have fun every day. Your brain will thank you.

APPENDIX A
The Age-Proof Brain Seven-Day Challenge

IF YOU WANT TO GET A JUMP START ON THE BRAIN-HEALTHY HABITS I outlined in part three, and you're up for a week of Brain Boot Camp, this weeklong challenge is one way to try things out and see what works best for you. Don't worry; this boot camp won't involve any bullhorns, obstacle courses, or treading water in jeans. This isn't designed to be a hell week. Rather, it's a fun week to explore new options and simple, small, practical changes.

Here's a full week's worth of activities, meal ideas, and suggestions to get you living the age-proof brain life! The goal of this challenge is not to have you do these same steps forever; rather, I want you to view this as a fun way to try out new things, see how you feel, and decide which practices you want to incorporate into your life moving forward after the seven days are up. This means that this week, you'll apply the same times for waking up and going to bed even on your days off from work or days when you don't *need* to be anywhere in the morning. (I know, I know—sleeping in on a Saturday is one of life's simple pleasures. But you may just discover that you enjoy those quiet morning hours, too.) If you see the wake-up time and freak out, you can adjust the schedule to fit your lifestyle.

Before we begin, a few caveats: this challenge accommodates a Monday to Friday, 9 AM to 5 PM work schedule. It also assumes you commute to work and your diet includes meat and dairy. (I know this won't apply to everyone!) If you are retired or semiretired, have a flexible schedule, work from home, work different hours, or are a vegetarian (or a vampire), see my suggestions on page 236 to customize the challenge.

I build in different times for a workout. Exercise timing is so personalized. Some people thrive with a morning workout, some like to get that boost in the afternoon, and some prefer the evening. This challenge allows you to experiment and see what works best for you. Of course, if you already know you just don't have the energy to exercise effectively after 5 PM, fit it in as you usually would. Also, a quick word on snacks. I include one suggested snack each afternoon. If you feel like you need another one, opt for something minimally processed, like nuts, fruit, hummus, veggies, or a healthy carb. See my tips at the end of the challenge for using the following as a template and adjusting it to suit your needs.

Ready to do this brain boot camp? Now drop and give me twenty push-ups. Just kidding, no need for that right now. Let's do this!

Monday

7:00 AM	Wake up!
7:15 AM	Get outside for a 10-minute walk.
7:30 AM	**Breakfast:** *Bananas and almond butter, by itself or on top of whole-grain bread for a brain supercharge. If you need more calories, try a hard-boiled egg and Greek yogurt with berries.*
8:00 AM	Time to head to work. Use your commute as learning time; listen to an informative podcast or audiobook.
9:00 AM	Workday start
12:00 PM	**Lunch:** *Wraps—take a whole grain tortilla and load it with leafy greens, steamed veggies, beans, and/or lean protein like tuna, salmon, tofu, or chicken.*
12:45 PM	10-minute after-lunch walk
3:30 PM	**Brain-healthy snack:** *Hummus and veggies*
5:00 PM	Reconnecting commute. Call a friend or relative you like to talk to.
6:00 PM	30-minute evening workout
6:45 PM	**Dinner:** *Frittata filled with a combination of broccoli, asparagus, mushrooms, spinach, roasted red pepper, and cherry tomatoes. Try feta or mozzarella mixed in as well. A side dish can be a simple salad or whole wheat roll.*
7:45 PM	10-minute after-dinner walk
9:30 PM	Device shut-off time. Try a little mindfulness before bed.
10:30 PM	Time for bed

Today's Tip: To get in the habit of going on that morning walk, put your walking shoes somewhere you will see them to remind you. Whether it's solo time or social is up to you; both have brain health benefits. Ask yourself what you need that day and take that time to get it.

Tuesday

7:00 AM	Wake up!
7:15 AM	Get outside for a 10-minute walk.
7:30 AM	**Breakfast:** *Oatmeal. Instant is fine; just avoid oatmeal with added sugar. Load it with berries (frozen or fresh work), and for a boost of protein, mix in almond butter. This is one of my favorites!*
8:00 AM	Time to head to work. Use your commute as learning time; listen to an informative podcast or audiobook.
9:00 AM	Workday start
11:30 AM	30-minute pre-lunch workout
12:00 PM	**Lunch:** *Amp up your favorite hearty sandwich. Opt for whole-grain bread and stay away from processed deli meat with nitrates. Add spinach, tomatoes, and/or avocado.*
12:45 PM	10-minute after-lunch walk
3:30 PM	**Brain-healthy snack:** *Beet chips*
5:00 PM	Comedy commute. Listen to some standup or a funny audiobook or podcast.
6:45 PM	**Dinner:** *Salmon with roasted root vegetables and a bit of olive oil and brown rice or quinoa. Make enough salmon, and maybe extra rice, so you'll have leftovers to enjoy tomorrow.*
7:45 PM	10-minute after-dinner walk
9:30 PM	Device shut-off time
10:30 PM	Time for bed

Today's Tip: If you have trouble focusing at work, try the Pomodoro method (described in chapter sixteen).

Wednesday

7:00 AM	Wake up!
7:15 AM	Get outside for a 10-minute walk.
7:30 AM	**Breakfast:** *Green smoothie. Blend up a combo that appeals to you of leafy greens, berries, banana, steamed beets, Greek yogurt, nut butter, and/or chia seeds, which are all brain-healthy options. Or, if you need something more substantial, try an egg on top of last night's leftover quinoa or brown rice.*
8:00 AM	Time to head to work. Use your commute as learning time; listen to an informative podcast or audiobook.
9:00 AM	Workday start
12:00 PM	**Lunch:** *Leftover salmon salad made from last night's cooked salmon on top of greens and veggies of your choice with vinaigrette. Add whole wheat bread if needed.*
12:45 PM	10-minute after-lunch walk
3:30 PM	**Brain-healthy snack:** *Walnuts and almonds*
5:00 PM	Arts and architecture commute. Or, science and history commute. Find an informative podcast on a subject you're interested in outside your career to listen to on your way home.
6:45 PM	**Dinner:** *Fajitas. Seasoned sauteed chicken or tofu, black beans, bell pepper strips, and onions with whole-grain tortillas. Save some of the protein and veggies for tomorrow's breakfast.*
7:45 PM	10-minute after-dinner walk
9:30 PM	Device shut-off time
10:30 PM	Time for bed

Today's Tip: Make eating healthy an easy choice by making treats something you have to leave your home to get. Keep fresh fruits and veggies front and center in your refrigerator and kitchen for easy snacking.

Thursday

6:30 AM	Wake up! (A bit earlier today if you're fitting in an early morning workout!)
6:45 AM	Get outside for a 10-minute walk. Use it as a warm-up, then go right into a 30-minute workout.
7:30 AM	**Breakfast:** *Leftovers scrambles. Use last night's healthy dinner components to add some extra flavor and texture to scrambled eggs. Try salsa on the side for a flavor boost. Add a healthy carb such as oatmeal or whole wheat toast if needed.*
8:00 AM	Time to head to work. Use your commute as learning time; listen to an informative podcast or audiobook.
9:00 AM	Workday start
12:00 PM	**Lunch:** *Chicken and avocado salad with blueberry balsamic dressing. Add a brain healthy carb such as brown rice or whole wheat pasta if needed.*
12:45 PM	10-minute after-lunch walk
3:30 PM	**Brain-healthy snack:** Cottage cheese with fruit or veggies
5:00 PM	Foreign language commute. If you speak or are learning a second language, try listening to the news or a podcast in that language. Or, try a beginner how-to in a new language that interests you. If the news or a podcast is too difficult, a great trick is to listen to preschool-age cartoon shows in a foreign language.
6:45 PM	**Dinner:** *Fish (like snapper) with miso and a side of steamed broccolini. If needed, add a healthy carb like brown rice, a whole wheat roll, or quinoa.*
7:45 PM	10-minute after-dinner walk
9:30 PM	Device shut-off time
10:30 PM	Time for bed

Today's Tip: Remember that it's often easier to fall asleep in a cool room. If you've been tossing and turning at night, revisit chapter ten and check your thermostat.

Friday

7:00 AM	Wake up!
7:15 AM	Get outside for a 10-minute walk.
7:30 AM	**Breakfast:** *Breakfast parfait. Layer unsweetened Greek yogurt with berries and nuts. Add poached eggs on whole grain toast with avocado if you need more fuel.*
8:00 AM	Time to head to work. Use your commute as learning time; listen to an informative podcast or audiobook.
9:00 AM	Workday start
11:30 AM	30-minute workout
12:00 PM	**Lunch:** *Sushi (salmon-avocado or cucumber rolls made with brown rice), edamame, and miso soup*
12:45 PM	10-minute after-lunch walk
3:30 PM	**Brain-healthy snack:** *Popcorn*
5:00 PM	TedTalk commute—listen to a TedTalk on the subject of happiness.
6:45 PM	**Dinner:** *Grilled chicken with sweet potatoes and steamed green beans. Add fruit for dessert.*
7:45 PM	10-minute after-dinner walk
9:30 PM	Device shut-off time
10:30 PM	Time for bed

Today's Tip: Though it's the kickoff to the weekend, resist the temptation to stay up later tonight. If you are absolutely exhausted from the week, you can shift a little bit and set your alarm a little later tomorrow. But ideally, keep your weekend bedtimes and wake-up calls the same as your work-week schedule; otherwise, it will throw off your internal clock for next Monday's wake up and bedtime.

Saturday

7:00 AM	Wake up!
7:15 AM	Get outside for a 10-minute walk.
7:30 AM	**Breakfast:** *Scrambled egg tacos. Load them with cooked or raw spinach, black beans, diced tomatoes, and guacamole.*
12:00 PM	**Lunch:** *Salmon with roasted vegetables. If needed, add a healthy carb such as brown rice, quinoa, or a whole wheat roll.*
12:45 PM	10-minute after-lunch walk.
1-3:00 PM	Try your 30-minute workout. Or, since it's Saturday, make it a longer one: Play a sport with some friends, try a new type of yoga or group fitness, or go for a challenging hike, bike ride, or canoe excursion.
3:30 PM	**Brain-healthy snack:** *Chia pudding.*
4:00 PM	Try learning a new skill. That might be doing an online tutorial for some sort of art/craft, learning a new song on your instrument, or even learning the latest TikTok dance. Just have fun with it–it's Saturday, after all.
6:45 PM	**Dinner:** *Lean pork chops with broccolini and zucchini noodles or whole-grain pasta.*
7:45 PM	10-minute after-dinner walk
9:30 PM	Device shut-off time
10:30 PM	Time for bed

Today's Tip: Take a social media sabbatical. At least take a couple of hours on the weekend and put your phone away; stay away from screens.

Sunday

7:00 AM	Wake up!
7:15 AM	Get outside for a 10-minute walk.
7:30 AM	**Breakfast:** *Veggie omelets. These are a great way to use up any leftovers you might have from last night's dinner. You can also try adding some salmon (fresh, canned, or leftover from lunch yesterday) for an added brain benefit.*
9:30–11:30 AM	Use this window for today's 30-minute workout.
11:30 AM	Put together tonight's dinner in the slow cooker.
12:00 PM	**Lunch:** *Stir-fry with lean meats and vegetables like peppers, carrots, cauliflower, mushrooms, and/or broccoli. Add brown rice or quinoa if needed.*
12:45 PM	10-minute after-lunch walk
3:30 PM	**Brain-healthy snack:** *Unsweetened Greek yogurt.*
6:45 PM	**Dinner:** *Slow cooker stew or roast. Stick to a recipe with lean meats and lots of veggies, and stay away from things that call for a lot of salt, heavy cream, or additives. If needed, add a healthy carb like a sweet potato.*
7:45 PM	10-minute after-dinner walk
9:30 PM	Device shut-off time
10:30 PM	Time for bed

Today's Tip: Today's challenge day has lots of open time. If you can, resist the urge to fill all of that time with a lot of errands, chores, and busy work. We live in a world where we feel like we always need to do something. Challenge yourself to do nothing—it is good for the brain. Close your eyes. Nap. Put your feet up. Go to a garden and look at the flowers. Watch the sun set. (I know these aren't really "nothing," but they are things we often don't give ourselves time to do.)

Customizing the Age-Proof Brain Seven-Day Challenge

Wake and sleep: Feel free to set different wake and sleep times if you need to leave for work earlier/later or if you know you don't need eight hours of sleep each night. If you're not sure how many hours of sleep you need, use the guide outlined in chapter ten to gauge it. Note that is a separate three- to four-day trial, with a two-hour "no devices" window before bed, and no alarms—so I don't recommend doing it during a time it might make you late for work.

Work schedule: If you don't work Monday through Friday, 9 AM to 5 PM, switch the work days to match your schedule. (I hope you get at least one day off to give your brain a break!) If you are retired, pick some times that work for you.

Creating a commute: If you work from home, or are lucky enough to live just a short distance from your workplace, create that same targeted brain time. Take twenty to thirty minutes in the morning and evening to do a "mental commute" that transitions you from work mode to relaxation. Or, take another walk, alone or with a buddy.

Adjusting for special diets: If you are vegan or vegetarian, swap in brain-healthy plant-based meals based on the guidelines in chapter fourteen. Follow the same basic principles: minimize additives and prioritize healthy fats and proteins, and make sure you are eating a rainbow of fruits and vegetables. Obviously, avoid any foods that trigger food sensitivities or allergies, and follow what works for you.

Squeezing in shorter workouts: I know how challenging it can be to fit 120 minutes of exercise into our busy lives. It's great to go to the gym, take an exercise class, or play a sport a couple of times a week, but if that doesn't fit into your day, here are some simple solutions that don't involve carving out a 30-minute window:

- Make one of your post-meal walks a longer run.
- Set aside three time windows during your work day for some express exercises.
- Try doing just 5 minutes of exercise every hour. In six or seven hours of sitting, you'll end up incorporating 30 minutes of exercise. Set a timer to go off every 55 minutes and do:
  - 1 minute of jumping jacks
  - 1 minute of lunges
  - 1 minute of squats
  - 1 minute of planks (yes, that's a long plank to start with—try working your way up!)
  - 1 minute of push-ups on the ground or against a wall

Hope you enjoy the challenge and find simple action steps to include in your day to age-proof your brain.

APPENDIX B
The One Important Sheet of Paper

THROUGHOUT THIS BOOK, I HIGHLIGHT SOME TESTS THAT YOU should ask about at your next routine physical exam. You may have been jotting them down on a sheet of paper to take with you. If you haven't, don't worry, I've got your back. You can make a copy of this page, snap a picture on your phone, or download a PDF of it at www.drmarcmilstein.com/ageproofbrain.

- **Lipid profile or lipid panel:** This blood test will measure your HDL, LDL, and total cholesterol levels. As discussed in chapter three, it's important to keep cholesterol within healthy levels (which vary depending on your age, sex, and medical history).
- **Homocysteine test:** This blood test will determine homocysteine levels, another important factor in protecting your cardiovascular system, as covered in chapter three. Healthy levels are usually between 5 and 15 micromoles per liter.
- **Hemoglobin A1C blood test (or glucose tolerance or fasting blood sugar test):** This test will determine whether you have prediabetes or type 2 diabetes, as we discussed in chapter seven.
- **C-reactive protein (CRP) test:** This blood test measures levels of inflammation. An elevated CRP has been linked to an increased risk of dementia, as covered in chapter eight.
- **Vitamins and nutrients:** Vitamin D, Vitamin B-12, and folate, as covered in chapter fourteen.

ACKNOWLEDGMENTS

I AM THANKFUL TO BE ABLE TO TAKE THIS OPPORTUNITY TO EXPRESS my gratitude and appreciation to so many who have helped and supported me along the way.

Thank you to my wife, Lauren, for being by my side every step of the way. I am incredibly lucky to have you as my partner and teammate. I cherish you and your endless support, conversations, and guidance in all aspects of life. Thank you for making every day better. Thank you to my daughters, Ella and Charlotte, for being a constant source of joy. I am so proud of both of you for the wonderful people you are and are becoming. I don't know how I am so fortunate to have you as my daughters. I am grateful every day.

Thank you to Howard and Barbara Milstein, for being extraordinary parents. A million thank yous. I admire you and I am so appreciative to be your son. Thank you for the thousands of conversations and lessons. Thank you to my sister, Rachel, for your constant support and for being not just a sibling but a best friend. I am so thankful and proud to have you as my sister.

I am very grateful for my supportive family: Ryan, Noah, and Lily Goldenhar. My thanks and admiration to Uncle Gabriel Wisdom for your positive and inspiring influence throughout my life. Thank you, Aunt Diana Weiss-Wisdom, for always being there. Thank you to my supportive family, Michael Alden, Alan and Vicki Schankman, Alison Schankman, Dana and Al Lincoln and family, Norman and Lorraine Milstein, Irv and Phyllis Wool, and Gertrude Anker Wool.

A heartfelt thank you to my literary agents, Margret McBride and Faye Atchison of the Margret McBride Agency, for their guidance and mentoring through this entire process. I am endlessly grateful and appreciative. Thank you for continually going above and beyond.

I am indebted to my editors at BenBella, Claire Schulz and Alyn Wallace. I cannot thank you enough for both elevating and preserving my ideas in this book. I thoroughly enjoyed working with you both. I am thankful for your creativity, attention to detail, talent, and fantastic ideas. I am so fortunate to have you as my editors.

A gigantic thank you to the entire BenBella team, including Glenn Yeffeth, Leah Wilson, Rachel Phares, Adrienne Lang, Jennifer Canzoneri, Sarah Avinger, Morgan Carr, Madeline Grigg, Alicia Kania, and Monica Lowry, for your hard work and tireless dedication in making this book come to fruition and for being an amazing and wonderful team to work with throughout the process.

Thank you to Sydny Miner for her editing on drafts of the proposal and drafts of this book. Your ideas, guidance, and editing immensely helped. Thank you to Neil Gordon and Ted Spiker for their editing and insights on early drafts of the proposal. Thank you to Judy Gelman Myers for your keen and thorough copyediting. Thank you all for being so excellent to work with.

I am beyond grateful and indebted to Dr. Arash Horizon and Dr. Eric Vasiliauskas for their care. Thank you to Dr. Jeanne Perry, Dayna Arnold, and Denis Waitley for their inspiration.

Thank you and much appreciation to Katie Hyde, Melissa Schroeder, Michaela Ciampi, Chris Marsh, Tammie Wallace, Esther Eagles, Jo Borello, Barbara Henricks, Alicia Underwood, and Robert Cima for helping facilitate an avenue to deliver the messages in this book. Thank you to lifelong friends Michael de la Cruz, Sang Kim, Kal Penn, David Frank, Michael Frank, and the members of Six South.

Finally, a huge, heartfelt thank you to all of those who have attended my series of talks in Los Angeles, San Diego, around the world, and virtually. It means the world to me that you have taken the time to attend the presentations. This book would not exist without you.

INDEX

ABOUT THE AUTHOR

Dr. Marc Milstein specializes in presenting the leading scientific research on brain health in a way that entertains, educates, and empowers his audience to live better. His presentations provide science-based solutions to keep the brain healthy, boost productivity, and maximize longevity. He earned a PhD in biological chemistry and a BS in molecular, cellular, and developmental biology from UCLA. Dr. Milstein has conducted research on topics including genetics, cancer biology, and neuroscience, and his work has been published in multiple scientific journals. He has been quoted in popular press such as *USA Today*, *HuffPost*, and *WeightWatchers* magazine. Dr. Milstein has also been featured on television explaining the latest scientific breakthroughs that improve our lives.

He lives in San Diego, California, with his wife, Lauren, and two daughters, Ella and Charlotte.